BIOLOGICAL INVESTIGATIONS II

LAB EXERCISES FOR GENERAL BIOLOGY II

Thirteenth Edition

Gretchen S. Bernard
and
Edward E. Devine
Moraine Valley Community College

Kendall Hunt
publishing company

www.kendallhunt.com
Send all inquiries to:
4050 Westmark Drive
Dubuque, IA 52004-1840

ISBN: 978-0-7575-8951-5

Printed in the United States of America
10 9 8 7 6 5 4 3 2 1

Name ______________________________ Section ______________

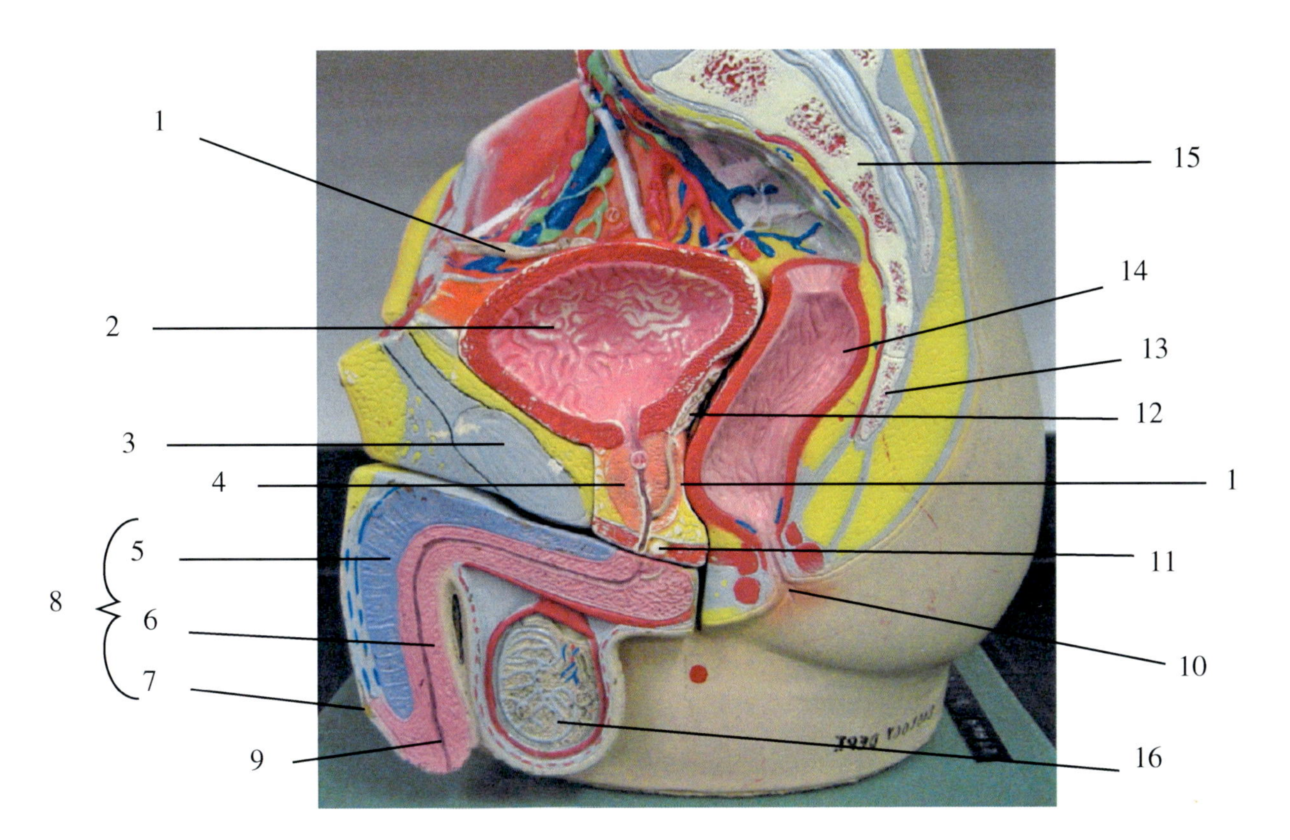

Figure 32.3
Male Urogenital System

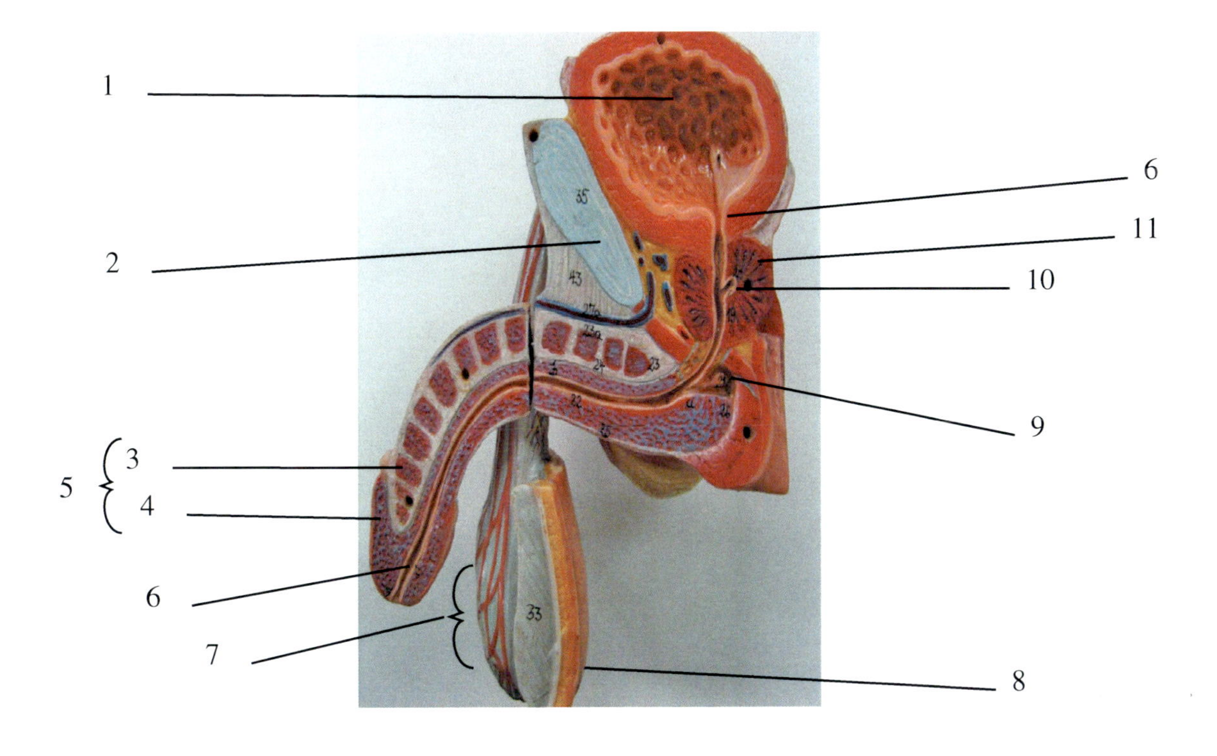

Figure 32.4
Male Urogenital System

Contents

Name ______________________________ Section ______________

Lab 21

Evidence of Evolution

Problem

Is there evidence that evolution has occurred?

Objectives

After completing this lab exercise, the student will be able to:

1. Compare chick and pig embryos at similar stages of development.
2. Distinguish homologous and analogous structures.
3. Identify the bones in the front limbs of an amphibian, reptile, bird, and mammal.
4. Name the transition animal between reptiles and birds.
5. List reptile characteristics and bird characteristics shared by their transition animal.
6. Note the changing position of the foremen magnum between vertebrate groups.
7. Arrange skulls of primates from primitive to modern.

Preliminary Information

Evolution is the change in the allele (gene) frequency of a population over a period of time. The evolutionary patterns fall into two major categories: microevolution and macroevolution. In **microevolution**, the changes are minor. They might involve body coloration, wing length, bone length, enzyme production, or other similar traits. In **macroevolution**, the changes are major, resulting in substantially different body shape and functioning so a new species may result.

In both microevolution and macroevolution, when a new species arises, it will retain many of the characteristics of the population from which it arose.

There is evidence of evolution from many areas indicating that different types of organisms are related to one another. In most instances, relatedness does not necessarily imply a direct evolutionary line from one type to the next but rather an indication that the types present now are related through some common ancestral stock.

This laboratory exercise will investigate some lines of evidence for evolution.

Name ______________________________ Section ______________

Part I: Evidence from Embryology

If organisms are related, their embryological development should be similar.

1. Obtain prepared microscope slides of:
   a. pig embryo
   b. chick embryo.
   These embryos are at comparable stages of development.
2. Using a **binocular microscope**, rotate the slide until the head is up, examine and diagram each embryo.
3. Label these parts on both:
   head (has 2 or 3 brain bulges)
   eye (dark, round)
   front limbs (small bulge near the heart)
   hind limbs (small bulge near the tail)
   heart (a prominent bulge near the middle)
   spinal cord
   tail

Vertebrate Embryos

Chick Embryo	Pig Embryo

Part II: Evidence from Limb Structure

If organisms are related, their limbs should have similar bones. **Homologous structures have similar shapes, development, and relationships with other surrounding structures.** The bones in the limbs of vertebrates are homologous structures. Analogous structures perform similar functions. A bird wing and a butterfly wing are analogous structures. **Homologous structures indicate relationships between organisms while analogous structures do not.**

1. Observe and diagram the forelimbs of the frog, turtle, chicken, bat, cat or goat, and human.
2. Label these bones on each diagram:
   humerus (first bone in the upper arm)
   radius (forearm bone on the thumb side)
   ulna (other bone in the forearm)
   carpals (wrist bones)
   metacarpals (long bones in the hand)
   phalanges (three finger bones).

Name ______________________________ Section ______________

Vetebrate Forelimbs

Frog	Turtle
Chicken	Bat
Cat or Goat	Human

Name ______________________________ Section ______________

Part III: Evidence of Evolution from a Transition Animal

A transition animal provides a line of relatedness.

The punctuated equilibrium theory of evolution states that the change from one species to another occurs very quickly in geological terms. The change happens so quickly that the transition forms are usually not found in the fossil record. The transition forms may not have been fossilized or the fossil remains may not have been discovered. **Transition forms** would certainly **provide good evidence of evolution**. *Archaeopteryx lithographica* is a transition animal between reptiles and birds.

1. Observe the fossil of *Archaeopteryx lithographica* at the demonstration area.
2. Complete the table below by **listing** some of the reptile characteristics and bird characteristics that this organism possesses.

Characteristics of *Archaeopteryx*

List Reptile Characteristics	List Bird Characteristics

Part IV: Evidence from Vestigial Structures and Lost Structures

Vestigial Structures

A vestigial structure shows no apparent purpose but is homologous to useful structures in other species. Examples of vestigial structures are hip bones in whales, hip bones in boas and pythons, the appendix is humans, and tail bones in humans.

1. Sketch the tail bones on the human skeleton in the space below.

Name ______________________ Section ____________

Loss of Limbs

Snakes are a recent group in reptile evolution. Snakes and legless lizards are the only groups of vertebrates that show an evolutionary loss of limbs. Snakes are thought to have evolved from lizards. Over time, most snakes have lost their limbs and supporting shoulder bones and hip bones. Boas and pythons retain remnants of their hip bones which can be seen protruding through the skin near the ventral posterior end of the snake. Due to loss of limbs, snake movement has reverted to primitive wiggling locomotion (undulations) seen in pre-fish vertebrates. Lateral undulations have been show to be an energy efficient means of locomotion requiring about half the energy of four-legged locomotion.

1. Examine a snake skeleton.
2. How many vertebrae does this snake have?

3. Do the ribs connect to the sternum?

4. Is there any evidence of shoulder bones or hip bones?

Part V: Evidence from Vertebrate Skull Structure

Similarities and differences in skulls show relatedness and adaptation to an upright posture.

Bipedalism (walking on two legs) is a basic marker of humans. The anatomy of the skull is an indicator of bipedalism. The skull of an upright walker is better balanced on the vertebral column with the foramen magnum being beneath the skull. The movement of the foramen magnum below the skull indicates a more upright walker.

1. Observe the indicated vertebrate skulls at the demonstration area.
2. **Diagram a side view of each of these skulls.**
   Note: Face all skulls the same direction.
3. Indicate the position of the foramen magnum (hole for spinal cord) on each skull with an arrow.

Name ______________________________ Section ______________

Position of Foramen Magnum

Turtle	**Dog or Fox**
Bird	**Chimp or Gorilla**
Australopithecus	**Human**

Name ______________________ Section ______________

4. Complete the table below to correlate some aspects of skull anatomy with posture. For each skull, note:
   a. eye orientation as:
      - sideways
      - forward
   b. position of foramen magnum as:
      - behind
      - angular
      - below
   c. posture as:
      - four-legged
      - two-legged, stooped
      - two-legged, upright.

Table 21.1
Some Aspects of Skull and Body

Animal	Eye Orientation	Foramen Magnum	Posture
Turtle			
Dog or Fox			
Bird			
Chimp or Gorilla			
Australopithecus			
Human			

Part VI: Evidence from Primate Skull Structure

This portion of the lab may be done as a class activity.

Anthropologists use many techniques in examining skulls to help them determine the degree of relatedness between animals. Some of these techniques involve complex mathematical measurements and comparisons involving mathematical data. They also use overall visual comparisons in their examinations.

In this portion of the lab, visual comparisons will be used to place the skulls in a line from primitive to modern. This does not necessarily imply that one group evolved from another but it can be used, along with other data, to show a degree of relatedness.

1. Examine the skulls of the monkeys, apes, hominid ancestors, and humans at the demonstration area.
2. Make a list of characteristics that can be used to distinguish and separate the skulls from one another.

Name ______________________________ Section ______________

3. List some Characteristics Used to Distinguish Skulls
   a.

   b.

   c.

   d.

   e.

4. Arrange the skulls in a linear order from least human-like to most human-like. All members of the class do not have to agree on the order. Use at least eight skulls.
5. Name the skulls in the order you have placed them. The names of the skulls are on the back of the skulls.

List the Arrangement of Primate Skulls

Name ______________________________ Section ______________

Part VII: Discussion

1. The embryos of all vertebrates go through many similar stages of development. What does this suggest?

2. Consider the human skeleton, the pig and chick embryo slides.
   a. Do adult humans have tail bones?

   b. What does this suggest?

   c. Would you expect a tail to be present in the human embryo if you examined one at the same stage of development as the chick and pig? Discuss these points..

3 a. Give examples of:
   - homologous structures.

   - analogous structures.

   - vestigial structures.

   b. Which of the above indicates a relationship between organisms?

4. As you examined the bones in the forelimbs of the various animals, what bones in the forelimbs changed:
   a. the most?

   b. the least?

Name ____________________ Section ____________

5. Select one animal you examined and explain how the shape of the bones in its forelimb is related to the type of movement or use.

6. Why is *Archaeopteryx* considered a good transition animal?

7. As animals changed from four-legged posture to a two-legged posture:
   a. what happened to the position of the foramen magnum?

   b. what is the advantage of the change in position of the foramen magnum?

8. As the overall picture of skull development from monkey to human is considered, discuss trends that appear regarding:
   a. shape of the face

   b. canine tooth structure

   c. bony projections on the skull

   d. strength of the lower jaw

   e. balance of the skull on the vertebral column.

Name ____________________ Section ____________

Lab 22

Bacterial Resistance to Antibiotics

Problem

Can bacteria become resistant to antibiotics?

Objectives

After completing this lab exercise, the student will be able to:

1. Demonstrate sterile technique when streaking and handling a bacterial culture plate.
2. Determine bacterial resistance to antibiotics.
3. Demonstrate the proper incubation procedure for a bacterial culture plate.
4. Discuss the relationship between mutations, environmental selection pressures, and evolution.

Preliminary Information

Evolution is a change in the gene frequency of a population over a period of time. To put it another way, evolution is the change in the percentage of one gene compared to its alternate in a population over many generations. **If the percentage of any characteristic in a population changes over time then the population has evolved.** Some changes in a population may go almost undetected while other changes may be quite dramatic. Each individual in a population has a set of genes that enables it to function in its particular environment. A mutation is an alteration in functioning genetic instructions. In most instances, this alteration will cause a detrimental effect when the genetic instructions stop functioning properly. However, in a very small percentage of mutations, the genetic change may be advantageous, making the individual better adapted to the environment, especially if the environment is changing in such a way that it selects for the mutation.

It is important to understand that:

1. an individual may undergo a mutation but it is the **population** that **evolves**.
2. **the environment always selects for preexisting characteristics in the population**. The selection factor can not cause the mutation.
3. a mutation is most often detrimental.

Bacteria are often used in experimentation because large populations can be grown in a small space, and a new generation can be produced every 20 minutes under optimum conditions. Bacteria do not need to mate to reproduce but simply divide in half by a process called binary fission.

In this experiment, the resistance of a population of bacteria to antibiotics will be tested.

Name ______________________________ Section ______________

Part I: Methods and Materials

Materials

The following will be needed for each team of two or as the instructor directs:

1 nutrient agar plate
2 safety glasses
1 permanent marker
1 forceps

Procedure

1. Wash and dry the top of the lab table. Keep it clear of all personal items.
2. Using a permanent marker and small printing, label the outer edge on the agar side of the plate with your name, date, and experiment.
3. Wear safety glasses.
4. Obtain a liquid culture tube of bacteria.
5. Using a sterile swab, stir the liquid bacterial culture.
6. Lift the top of the petri plate and, with the swab, spread the bacteria evenly over the agar. Rotate the plate 90 degrees (a quarter turn) and repeat streaking. Be sure the entire surface is covered with bacteria.

7. Return swab to the bacterial tube.
8. Pass the bacteria tube and swab on to the next table.
9. The instructor will dispense the antibiotics by placing the antibiotic disc dispenser over the opened petri plate. Each disc has the antibiotic name and concentration printed on it.
10. Use forceps to arrange antibiotic discs at equal distances around the petri plate as needed.
11. Immediately place forceps into discard beaker in the supply area.
12. Cover the petri plate.
13. Let the plates sit right-side-up for 10 minutes so the discs make firm contact with the agar.
14. Place the bacterial plates, up-side-down in the tray for your class in the incubator at 37° Celsius for 2 days.
    Note: Plates must be cultured and stored up-side-down so water does not condense on the bacteria side of the plate.
15. Sterilize the lab table, using alcohol and paper towels.
16. Wash your hands.

Name ____________________ Section ____________

Part II: Data

1. Record the antibiotics and concentrations you used in the data table.
2. Examine four plates and record whether the bacteria are:
   - ❖ sensitive (clear zone around antibiotic)
   - ❖ intermediate (halo or spotty colonies in clear zone)
   - ❖ resistant (growth up to antibiotic).
3. In the table, also record (in parentheses) the number of colonies found in the clear areas surrounding any antibiotic disc. **A colony in a clear zone indicates evolution is occuring; a bacteria (and its descendents) has mutated, resulting in some resistance to the antibiotic.**
4. **Remember the antibiotic did not cause the mutation, it only selected for those bacteria that already possessed the mutation for resistance to a specific antibiotic.**

Table 22.1
Antibiotic Disc Symbols

Symbol	Antibiotic	Concentration
AM	Ampicillin	10 ug
C	Choloramphenicol	30 ug
E	Erythromycin	15 ug
K	Kanamycin	30 ug
N	Neomycin	30 ug
P	Penicillin	10 units

Name ______________________ Section ______________

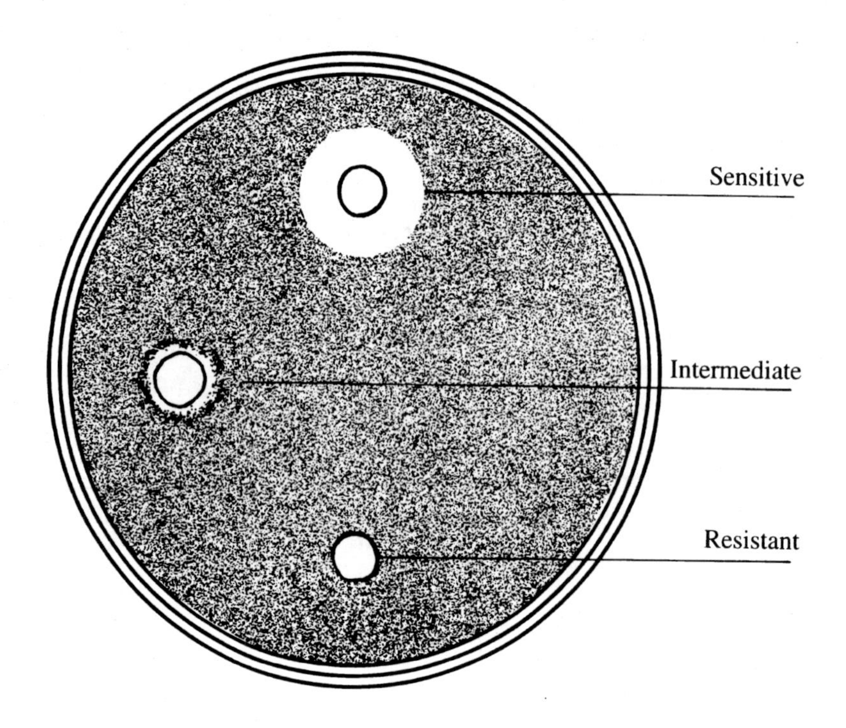

Figure 22.1
Bacterial Growth with Antibiotic

Table 22.2
Antibiotic Effectiveness

Antibiotic/ Concentration	Bacterial Sensitivity			
	Plate #1	Plate #2	Plate #3	Plate #4

Name ______________________________ Section ______________

Part III: Discussion

1. Name the antibiotics to which the bacteria are:
   a. completely resistant.

   b. intermediate resistant.

2. List the antibiotics from most to least effective on this strain of bacteria.

3. If penicillin resistance appears in the bacteria population did the penicillin cause the mutation? Explain.

4. Bacteria did not originally possess antibiotic resistance. Explain how some bacteria now have resistance to some antibiotics?

Name ______________________ Section ______________

Lab 23

Hardy-Weinberg Calculations

Problem

Can the percentages of dominant and recessive genes in a population be determined?

Objectives

After completing this lab exercise, the student will be able to:

1. Explain the significance of the Hardy-Weinberg Equation.
2. State the conditions that must be present for the allele frequencies to remain constant from generation to generation.
3. Calculate the percent of dominant and recessive genes in a population from the phenotypic ratios.
4. Predict the phenotypic and genotypic ratios in successive generations when the starting generation is known.

Preliminary Information

In the early 1900s, the field of population genetics began to be studied. In 1908, G. H. Hardy, an English mathematician, and G. Weinberg, a German physician, developed independently and published almost simultaneously a mathematical explanation of why certain genetic traits persist in a population generation after generation. Their work shows that dominant genes do not replace recessive genes. Their mathematical expression is called the Hardy-Weinberg Equation. According to the **Hardy-Weinberg Equation**, the percentages of dominant and recessive genes in a population will remain the same generation after generation if these conditions are met:

1. Large population
2. Isolated population
3. Random mating
4. No mutation
5. No natural selection.

If these conditions are met, then the percent of genes in a population cannot change and evolution cannot occur. **Evolution is the change in the allele (gene) frequencies of a population over a period of time.** However, if the percentages of genes are not the same generation after generation, then evolution has occurred.

Name ______________________________ Section ______________

The Hardy-Weinberg Equation can be used when there are only two alleles involved for one trait. One example of the use of the Hardy-Weinberg Equation can be seen in a fruit fly population. The equation can begin with the percent of each type of fruit fly present in the experimental population. For example, if a population is set up with seven homozygous dominant wild-type (gray body) flies and three homozygous recessive ebony-body flies, then 70% of the population has gray bodies and 30% of the population has ebony bodies. This can be expressed mathematically like this:

Table 23.1
Proportion to Percent

Trait	Proportion	Percent	Frequency
Wild	7/10	70%	.7
Ebony	3/10	30%	.3

The Decimal Equivalent is Used in Hardy-Weinberg Calculations.

If these flies breed randomly, the next generation can be calculated by either of two methods: the Hardy-Weinberg square method or the binomial expansion method. In either of these methods "p" represents the frequency of dominant alleles in the population for a particular trait and "q" represents the frequency of recessive alleles in the population for the same trait. Since there are only two alleles for a trait, then:

$p + q = 1.$

In other words, the percentage of dominant alleles for a trait plus the percentage of recessive alleles for the same trait in a population equal 100% of the possible alleles for the trait.

Note: All calculations must be done in the decimal form, not percent form. The decimal form is also called the frequency.

Hardy-Weinberg Square Method

	p	q
p	pp	pq
q	pq	qq

now add frequency

	.7	.3
.7	.49	.21
.3	.21	.09

Binomial Expansion Method

$$\begin{array}{r} p + q \\ \text{X}\ \underline{p + q} \\ pp + pq \\ \underline{pq + qq} \\ pp + 2pq + qq \end{array}$$

now add frequency

$$\begin{array}{r} .7 + .3 \\ \text{X}\ \underline{.7 + .3} \\ .49 + .21 \\ \underline{.21 + .09} \\ .49 + .42 + .09 \end{array}$$

or $p^2 + 2pq + q^2$

The interpretation of either the Hardy-Weinberg square method or the binomial expansion method is the same:

- 49% of the population is homozygous dominant
- 42% of the population is heterozygous
- 9% of the population is homozygous recessive.

Name ______________________________ Section ______________

Examining a Population

1. When you come upon a population, there is no way to know what the parents were. Therefore, a round-about way must be used to determine the percent of genes present in the population.
2. If the alleles for a characteristic are dominant-recessive, there is no way to distinguish homozygous dominant and heterozygous since they appear the same. Therefore, **the beginning point in examining the allele frequency of a population is the homozygous recessive**. When the percent of homozygous recessive is known, it is possible to work backward to calculate the percent of recessive alleles in the population, and from there the percent of dominant alleles.
3. In the Hardy-Weinberg Equation, the **percent of genotypes** in the population is represented by the equation:

   $p^2 + 2pq + q^2 = 1$ or
   Homozygous dominant + Heterozygous + Homozygous recessive = 100% of the possibilities for the trait
4. The **percent of alleles** in the population is represented by the equation:
   $p + q = 1$ or
   Dominant alleles + recessive alleles = 100% of the alleles in the population for the trait
5. Now examine the fruit fly numbers using the Hardy-Weinberg Equation. It says that 9% of the population (q^2) will show the recessive trait. However, since many of the recessive alleles are present in the heterozygotes, the total percentage of recessive alleles (q) in the population can be calculated by taking the square root of q^2. In this case:

   if $q^2 = .09$
   then $q = .3$

which means that 30% of the alleles in this population are recessive.

Once the frequency of the recessive alleles has been calculated, the frequency of the dominant alleles can be calculated by the formula:

$$p = 1 - q$$

In this case,

$$p = 1 - .3$$
$$p = .7$$

This means 70% of the alleles in a population are dominant.

Name ______________________________ Section ______________

6. Once the allele frequencies are known, the Hardy-Weinberg square can be completed.

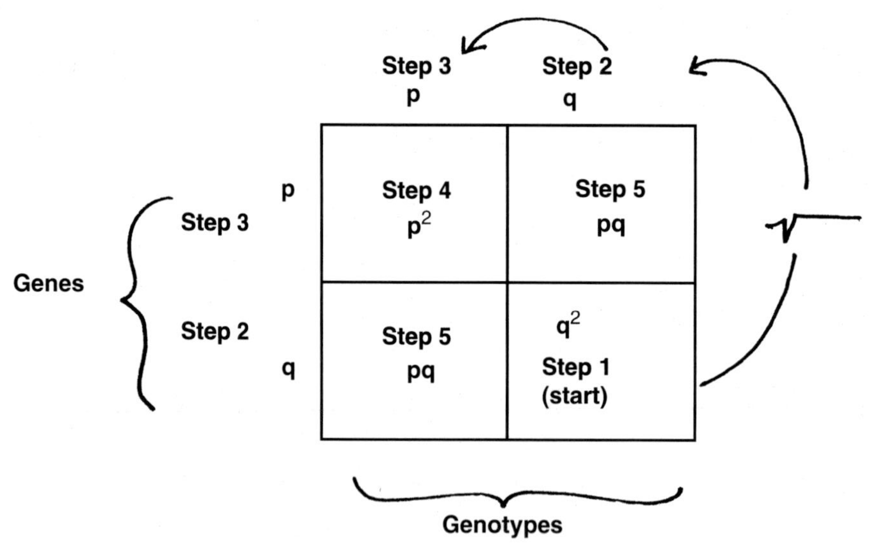

Figure 23.1
Steps to Examining a Population

Table 23.2
Hardy-Weinberg Summary

Frequency of Genotypes	$p^2 + 2pq + q^2 = 1$
Frequency of Alleles	$p + q = 1$

Part I: Examination of a Human Population

Humans possess a few characteristics with a dominant-recessive relationship. In this portion of the lab, the class will be the population.

Right- and Left-Handedness

1. How many students are in the class? ______________
2. In humans, right-handedness is dominant over left-handedness. How many left-handed people are in the class? ______________
3. What is the frequency of left-handedness in the class? ______________
   This is q^2. Write it in the Hardy-Weinberg square.
4. Calculate q by taking the square root of q^2.
   Put q on the Hardy-Weinberg square.
5. Calculate p ($p = 1 - q$). Put p on the Hardy-Weinberg square.
6. Fill in the Hardy-Weinberg square.
7. Summarize the data in the table.

Name ______________________ Section ____________

Table 23.3
Percent of Right- and Left-Handedness

% Homozygous Right	% Heterozygous Right	% Homozygous Left
% Dominant Alleles	% Recessive Alleles	
% Appearing Dominant	% Appearing Recessive	

Earlobes

1. How many students are in the class? ____________
2. In humans, free earlobes are dominant over attached. How many people have attached earlobes? ____________
3. What is the frequency of attached earlobes in the class? ____________

   This is q^2. Write it in the Hardy-Weinberg square.
4. Calculate q by taking the square root of q^2. Put q on the Hardy-Weinberg square.
5. Calculate p ($p = 1 - q$). Put p on the Hardy-Weinberg square.
6. Fill in the Hardy-Weinberg square.
7. Summarize the data in the table.

Table 23.4
Percent of Earlobes

% Homozygous Free	% Heterozygous Free	% Homozygous Attached
% Dominant Alleles	% Recessive Alleles	
% Appearing Dominant	% Appearing Recessive	

Name ______________________________ Section ______________

Hairline

1. How many students are in the class? ______________
2. In humans, widow's peak is dominant over straight hairline across the forehead. How many people have straight hairline? ______________
3. What is the frequency of straight hairline in the class? ______________

   This is q^2. Write it in the Hardy-Weinberg square.
4. Calculate q by taking the square root of q^2. Put q on the Hardy-Weinberg square.
5. Calculate p ($p = 1 - q$). Put p on the Hardy-Weinberg square.
6. Fill in the Hardy-Weinberg square.
7. Summarize the data in the table.

Table 23.5
Percent of Hairlines

% Homozygous Widow Peak	% Heterozygous Widow Peak	% Homozygous Straight
% Dominant Alleles	**% Recessive Alleles**	
% Appearing Dominant	**% Appearing Recessive**	

Part II: Predicting Later Generations

1. Select one of the traits examined in Part I. What trait do you want to use? ______________
2. The members of this class will be the parent generation.
3. Use the data from Part I. What **percent of the alleles** in this class population are:

   dominant (p)? ______________

   recessive (q)? ______________
4. Hypothetically mate the members of this class to produce the next generation (F_1) by filling in the Hardy-Weinberg square with frequencies of the percentages from above:

	p ____	q ____
p ____		
q ____		

Name ______________________________ Section ____________

5. In the F_1 generation, what **percent** are:

homozygous dominant? ____________

heterozygous? ____________

homozygous recessive? ____________

6. In the F_1 population, what is the **frequency** of:

dominant alleles (p)? ____________

recessive alleles (q)? ____________

7. Hypothetically mate the members of the F_1 generation to obtain the F_2 generation by filling in the Hardy-Weinberg square:

	p ____	q ____
p ____		
q ____		

8. In the F_2 generation, what **percent** are:

homozygous dominant? ____________

heterozygous? ____________

homozygous recessive? ____________

9. In the F_2 population, what is the **frequency** of:

dominant alleles (p)? ____________

recessive alleles (q)? ____________

10. Does the allele frequency change in this population from generation to generation? Discuss.

Name______________________________ Section______________

Part III: Discussion

1. What conditions must be present for the percentages of alleles in a population to remain constant from generation to generation?

2. As you examine a population over several generations, how can you tell if a population is evolving?

3. Use your own words to explain the meaning of:
   a. p

   b. q

   c. p^2

   d. 2pq

   e. q^2

4. When you examine a population using the Hardy-Weinberg Equation, why must you begin with the percentage of individuals with the recessive trait rather than those with the dominant trait?

Name ______________________________ Section ______________

5. A recessive trait is seen in 4% of a population. Using this as a starting point, complete the Hardy-Weinberg square and the table below. Then answer the associated questions.

Table 23.6
Hardy-Weinberg Population Analysis

p	q	p^2	2pq	q^2

a. What percent of the **population** is homozygous dominant? ______________

b. What percent of the **population** is heterozygous? ______________

c. What percent of the **alleles** in the population are dominant? ______________

d. What percent of **alleles** in the population are recessive? ______________

e. In the next (F_2) generation, what percent of individuals will show the recession trait?
____

f. In the next (F_2) generation, what percent of the individuals will show the dominant trait?
____

6. In humans, freckles is a dominant trait. 64% of a population has freckles.
a. What percent of the **population** has the recessive trait (no freckles)? ______________

b. Use this data to complete the following Hardy-Weinberg square.

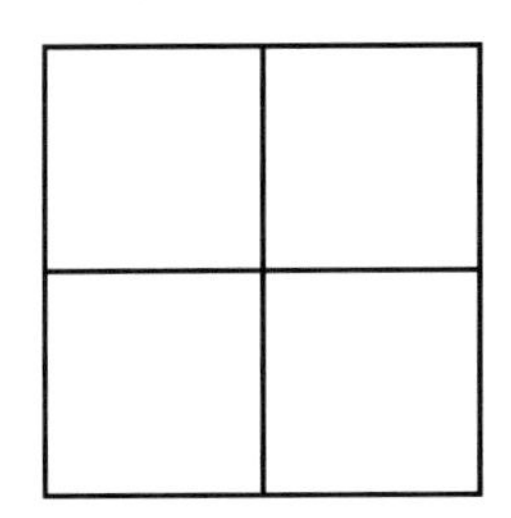

c. What percent of the **population** is homozygous dominant? ______________

d. What percent of the **population** is heterozygous?
____

e. What percent of the **alleles** in the population are dominant?
____

f. What percent of **alleles** in the population are recessive? ______________

g. In the next (F_2) generation, what percent of individuals will show the recession trait?
____

h. In the next (F_2) generation, what percent of the individuals will show the dominant trait?
____

7. In fruit flies, gray body color is dominant over ebony body color. In a gene pool of fruit flies, 84% of the alleles for body color are dominant.
Show all calculations.
   a. What is the frequency of the **dominant allele**? ______________
   b. What is the frequency of the **recessive allele**? ______________
   c. What percent of the **population** is homozygous dominant? ______________
   d. What percent of the **population** is heterozygous? ______________
   e. What percent of the **population** is homozygous recessive? ______________

8. In fruit flies, long wing is dominant over vestigial wing. A researcher examining a population of 250 fruit flies finds that 27 of them have vestigial wings.
Show all calculations.
   a. What percent of the **population** shows the recessive trait? ______________
   b. What percent of the **population** shows the dominant trait? ______________
   c. What percent of the **alleles** in the population are recessive? ______________
   d. What percent of the **alleles** in the population are dominant? ______________
   e. What percent of the **population** is homozygous dominant? ______________
   f. What percent of the **population** is heterozygous? ______________
   g. What percent of the **population** is homozygous recessive? ______________

Name ______________________________ Section ______________

Lab 24

Effect of Hormone on Plant Growth

Note: The seeds germinate better if soaked for 60-80 minutes before planting.

Problem

What effect will gibberellins have on plant growth in normal and dwarf pea plants?

Objectives

After completing this experiment, the student will be able to:

1. Identify the role of a hormone in plants.
2. Discuss the effect gibberellic acid has on the growth of normal and dwarf pea plant stems.
3. Construct a table and a graph from the data obtained.
4. Explain the use of a control in an experiment.
5. Compare the germination rates between normal and dwarf pea plants.

Preliminary Information

Maintenance and growth in plants is under the regulation of plant hormones. Like animal hormones, plant hormones are synthesized in one part of the organism but have an effect in a different part of the organism. Hormones are effective in minute quantities. Gibberellin or gibberellic acid is one type of plant hormone.

Gibberellins specifically promote cell enlargement but do not cause curvatures of the stem. The most common effect attributed to application of gibberellins is the extreme elongation of the stem, but it is also involved in flowering response and seed sprouting (**germination**).

Hypothesis

In this experiment gibberellins will be applied to both dwarf and normal pea plants. Before beginning the experiment it is best to give some thought to how you expect the experiment to turn out. Use the preliminary information and develop a probable outcome or hypothesis. If in the end, the hypothesis is found to be incorrect, this in no way decreases the validity of the hypothesis.

Hypothesis: ____

____

Name ____________________ Section ____________

Part I: Stem Growth

Materials

The following will be needed for each team of four or as the instructor directs:

20 presoaked normal pea seeds
20 presoaked dwarf pea seeds
1 planting tray with soil
Plastic wrap
Masking tape

Note: Seeds should be presoaked for 60-80 minutes.

Divide the tray into fourths with strips of masking tape and label as follows:

Normal control	Normal treated
Dwarf control	Dwarf treated

Record all names and the section number on tape and apply to the tray.

Procedure

1. Label the tray as suggested under materials section.
2. Plant 10 normal pea seeds in each section labeled "normal control" and "normal treated." Plant 10 dwarf pea seeds in each section labeled "dwarf control" and "dwarf treated."
3. Poke holes one cm deep with a pencil. Place seeds in the holes. Cover seeds with soil.
4. Add enough water to moisten all of the soil.
5. Check daily that plants are moist enough. **The soil should feel damp to the touch but not wet.** Overwatering or underwatering can affect how many seeds will germinate.
6. Particularly make sure all plants are watered on Friday afternoon and Monday morning.
7. Cover trays loosely with plastic wrap and tack each side with masking tape. Fold the plastic wrap back at alternate corners to leave a 2-3 cm opening for air circulation. Remove the plastic when plants are 2-4 cm high.
8. Place trays under plant growth lights.
9. Seven days after planting:
   a. Number each plant for identification with a small durable paper collar or flags made of tape and toothpicks.
   b. Measure the height in millimeters from the surface of the soil to the tip of the shoot of each plant.
   c. Record all initial measurements on data sheets under Day 0.
10. Spray the treated normal and treated dwarf peas with gibberellic acid. Spray the leaves and apex until droplets form. Avoid overspraying. (***Caution.*** Cover control peas so they are protected from the spray.)
11. Measure all plants and spray treated plants during each class for 2 weeks. Record all data.
12. Calculate and record the percent of increase for each group for each day (formula is with data sheet).
13. Graph the effect of gibberellic acid on stem growth by plotting time in days on the horizontal axis and percent of increase in length on the vertical axis.

Name ____________________ Section ____________

Part II: Data

Table 24.1
Effects of Gibberellic Acid on Stem Growth

Type	Plant Number	Initial Length (mm) Day 0	Length in mm				
			Day ____	Day ____	Day ____	Day ____	Day ____
Normal control	1						
	2						
	3						
	4						
	5						
	6						
	7						
	8						
	9						
	10						
	average						
	% increase	//////					
Dwarf control	1						
	2						
	3						
	4						
	5						
	6						
	7						
	8						
	9						
	10						
	average						
	% increase	//////					

Note: Always record the number of days that have passed *since* the initial length measurement was taken.

$$\% \text{ Increase for Day 2} = \frac{\text{Average length Day 2—Average initial length}}{\text{Average initial length}} \times 100$$

Use "Average initial length" for all calculations, not the average from the previous day.

Repeat % Increase calculations for each day data was recorded.

Name ______________________________ Section ______________

TABLE 24.1 (continued)

Type	Plant Number	Initial Length (mm) Day 0	Length in mm				
			Day _____	Day _____	Day _____	Day _____	Day _____
Normal treated	1						
	2						
	3						
	4						
	5						
	6						
	7						
	8						
	9						
	10						
	average						
	% increase						
Dwarf treated	1						
	2						
	3						
	4						
	5						
	6						
	7						
	8						
	9						
	10						
	average						
	% increase						

Name ______________________________ Section ______________

Part III: Graph Effect of Gibberellic Acid

1. Before beginning graph construction, calculate the percent of increase of length for each day. The graph should be constructed on a separate sheet of paper. Remember to include a **title** and **label** each axis.
2. Plot all curves on the same graph and **label each curve**.
3. Construct a line graph.
4. Plot **time (days) on the horizontal axis and percent of increase of length on the vertical axis**. In plotting time, remember to include days over the weekend even though no measurements were taken on these days.

Notes on Graph Construction

1. A **graph** is a **picture** of the data.
   a. **Plan** the graph to properly fit and fill the page.
   b. Do not write to the edge of the page. A minimum of 2.5 cm (1 inch) **margin** must be on all sides.
   c. Allow space for numbers, labels, title, and margins before placing each axis.
2. The factor under the experimenter's control (usually time) is placed on the horizontal axis. The measured factor is placed on the vertical axis.
3. **Intervals** on each axis should be **equal** and appropriate.
4. Use a **straight edge** when drawing each axis and connecting data points (don't be sloppy).
5. A graph should be self-explanatory:
   a. **Title** must be present.
   b. Completely **label each axis** with factor, units, scale, etc. For example: "Temperature" is not complete but "Temperature °C" is complete.
   c. Label each curve, if the graph has more than one.

Part IV: Discussion

1. Why bother with the control plants if they are not going to be sprayed with gibberellic acid?

2. Do the normal pea plants respond to the treatment of gibberellins in the same way as the dwarf plants? Why or why not?

3. In your own words, define **germination**.

Name ______________________________ Section ______________

4. Calculate percentage of germination for normal and dwarf plants.

$$\frac{\text{Number of seeds germinated}}{\text{Number of seeds planted}} \times 100 = \text{Percent of germination}$$

Percent of germination for normal plants (combine normal control and normal treated) =

____

Percent of germination for dwarf plants (combine normal control and normal treated) =

____

5. Did you observe any difference in **germination rates** between normal and dwarf plants? What is the significance of this?

6. What are the functions of gibberellins in a normal plant?

7. From your data, what can you conclude as to the probable cause of dwarf plants?

Name ____________________ Section ____________________

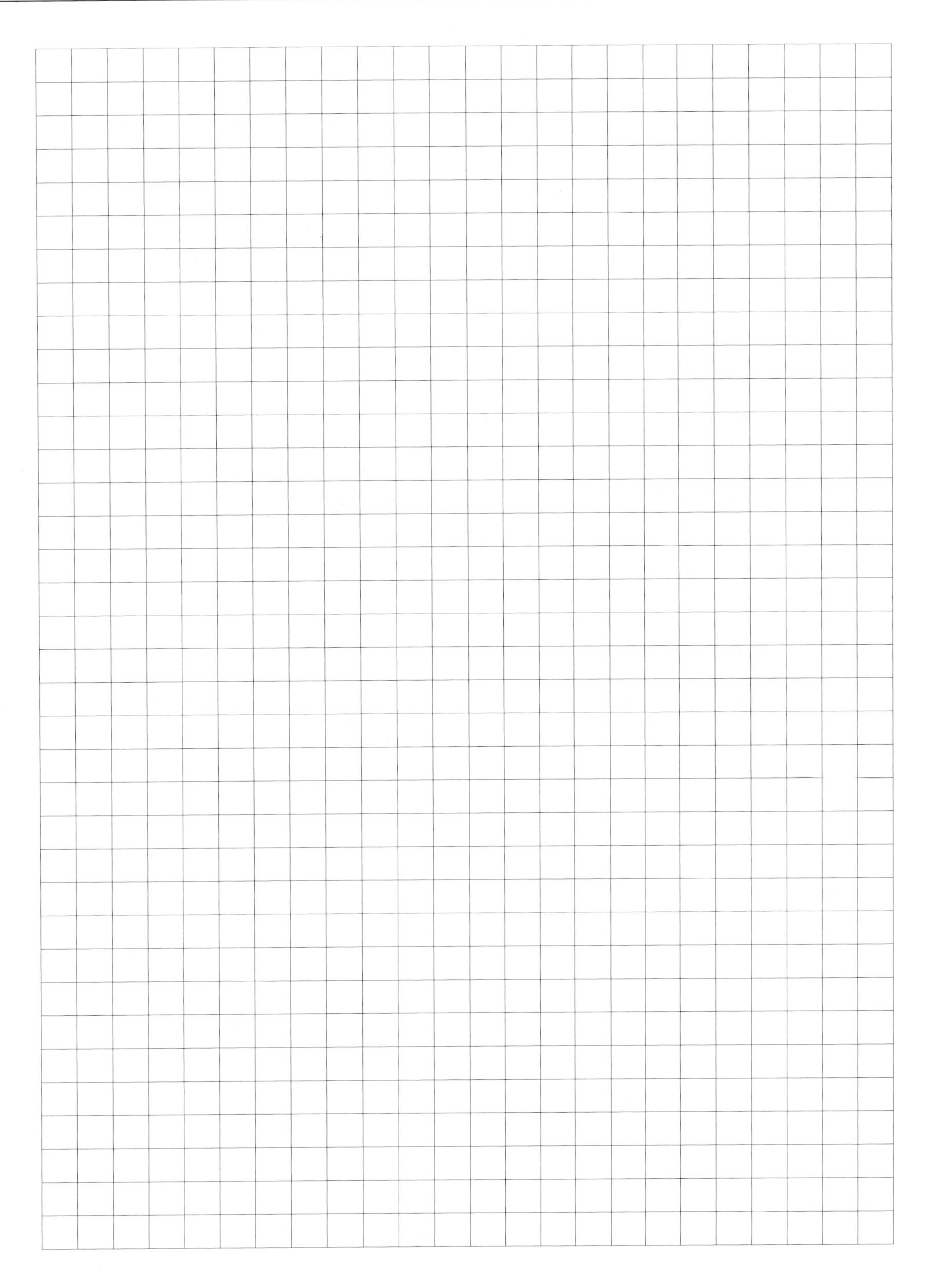

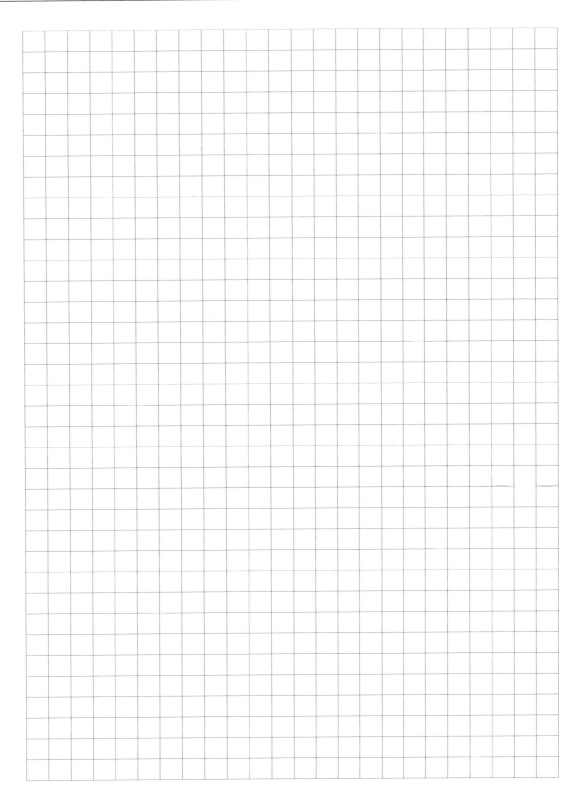

Name ______________________ Section ______________

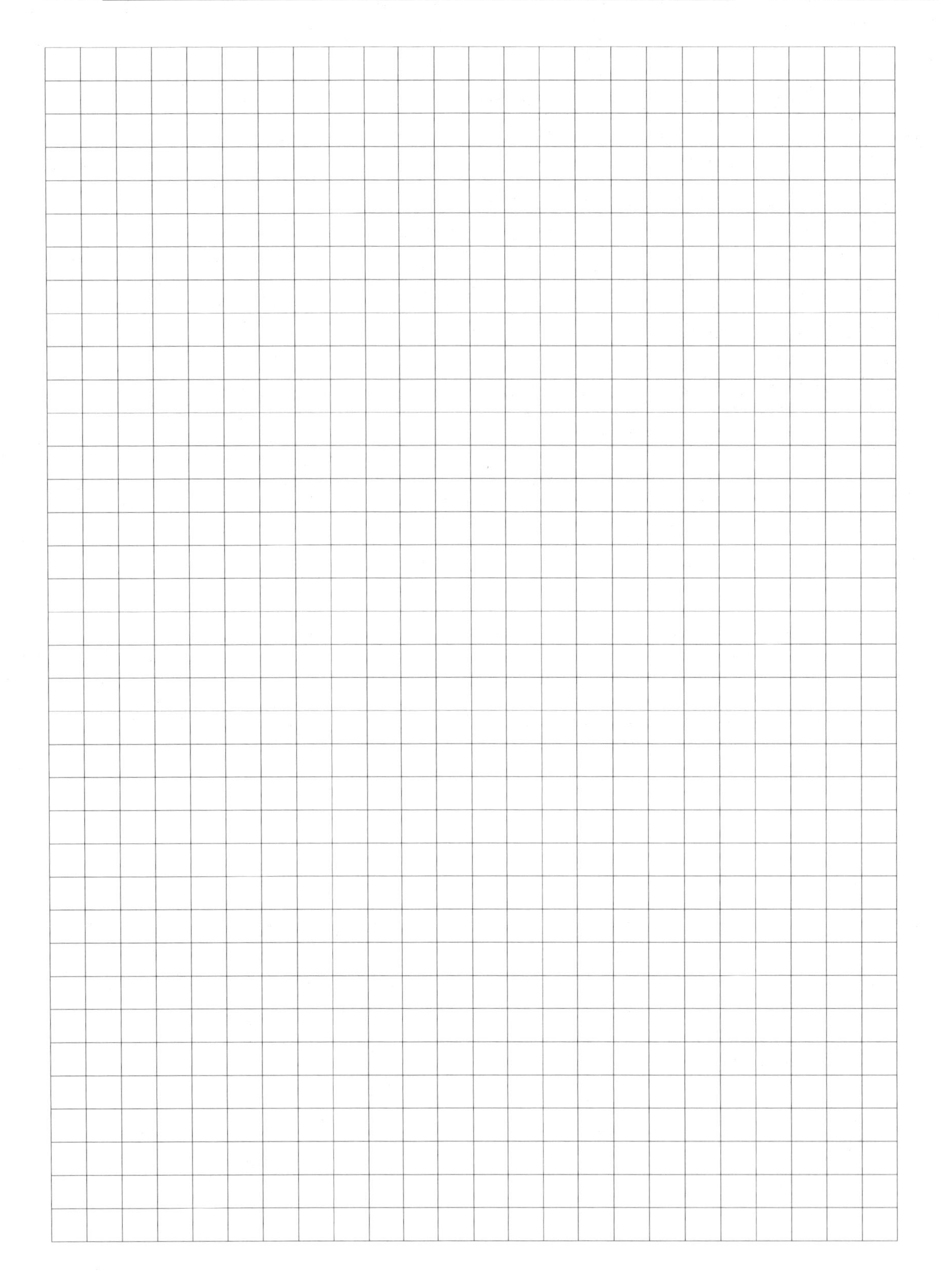

Name ____________________ Section ____________

Lab 25

Chlorophyll Extraction and Separation

Problem

Can the presence and function of plant pigments be determined?

Objectives

After completing the exercise, the student will be able to:

1. Discuss the method used to extract plant pigments.
2. Identify the number of pigments found in a leaf.
3. Explain the process of paper chromatography.
4. Relate the color of plant pigments in a leaf to the process of photosynthesis.
5. Identify a simple test to detect the presence of starch.

Preliminary Information

The most predominant plant pigment is chlorophyll. Chlorophyll is necessary for photosynthesis. It gives the leaf its green color because certain wavelengths of light are absorbed and green wavelengths are reflected by the plant. Our eyes see this reflected green color.

Chlorophyll often masks the presence of other plant pigments. In the autumn, the chlorophyll is no longer produced and other plant pigments can be seen. This is why the marvelous array of reds and yellows appear in the fall. These pigments can be seen in plants with variegated leaves. Variegated leaves have patches of white, yellow, or pink, which lack the green pigment, chlorophyll.

Photosynthesis is the biochemical process that results in the production of a simple sugar. Some of the simple sugar produced in the leaf is converted into the more complex form of starch. The presence of starch can be tested by adding iodine. The starch will appear brown or purple in the leaf.

Paper chromatography is a technique that can be used to separate plant pigments. This process involves dissolving the pigments in a solvent then allowing the pigments to move up a strip of chromatography paper with the aid of a second solvent. Since the pigments are of different solubilities, they move up the paper at different speeds and can thus be separated from each other.

Name ____________________ Section ____________

Part I: Pigment Extraction and Separation

Materials

Each group should obtain the following materials from the supply area:

One strip of chromatography paper. Handle by the edges only.
Scissors
One large test tube.
One cork or rubber stopper with wire hook to fit test tube.
Note: The wire hook will be used to suspend the chromatography paper inside the test tube. See Figure 25.1.
Fine tipped pipette.
Mortar and pestle.
One spinach leaf.
Approximately 6 mL of acetone in the mortar.

Note: One dropper full of acetone is about 1 mL.

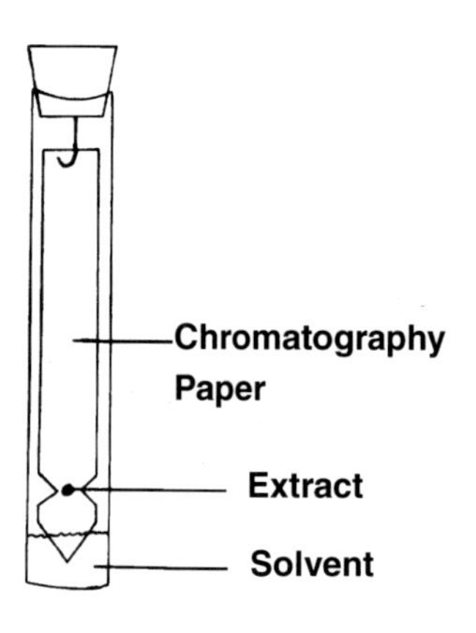

Figure 25.1
Chromatography Apparatus

Procedure

Note: Wear safety glasses.
Note: Refer to Figure 25.1 for the assembled chromatography apparatus.

1. Do not touch the face of the chromatography paper.
   Handle it by the edges.
   Trim the chromatography paper to fit into the test tube. When the paper is suspended, it should hang straight without touching the sides or bottom of the test tube.
   Trim the bottom of the chromatography paper to a V-shaped point. Approximately 2 cm from the bottom, cut a V-shaped notch out of each side of the chromatography strip.
2. Prepare a chlorophyll extract by thoroughly grinding spinach leaves in approximately 6 mL of acetone using the mortar and pestle. Cover with a paper towel to prevent splashing. **Note: A full squeeze from a dropper bottle equals about 1 mL.**
3. Using a fine tipped pipette, apply a VERY THIN BAND or a DROP of extract between or slightly below the notches in the chromatography paper.
4. **Allow the extract to dry completely.** Blowing the paper will help it dry faster.

5. Repeat steps 3 and 4 until 10 coats of extract have been applied.
6. Add 3 mL of petroleum ether-acetone solvent to the test tube.
7. Attach the strip of chromatography paper to the stopper and place it in a test tube. **Caution:** The band of extract must not be submerged in the solvent. Place the tip of the paper 1 cm into the solvent. The paper must hang straight and not touch the sides of the test tube.
8. Carefully place the test tube in a large beaker containing some crumpled paper towels to hold the test tube straight.
9. Observe at about 3-minute intervals until four color bands are visible. When a thin bright yellow band reaches the top of the strip, remove the strip to a paper towel to dry.
10. Save the chromatography strip for later use.
11. Dispose of the petroleum ether-acetone solvent in the marked waste container.

Part II: Pigments Responsible for Photosynthesis

Materials

Each group should obtain the following materials from the supply area:

One variegated leaf
500-mL or larger beaker
Large test tube
Alcohol, fill test tube half full
Petri dish
Iodine in dropper bottle
Hot plate

Procedure

Note: Wear safety glasses.

1. Using a hot plate, begin heating half a large beaker of water to boiling.
2. Obtain a large test tube and fill it half full of alcohol.
3. Add the leaf to the alcohol.
4. Place alcohol test tube with leaf in the beaker of lightly boiling water.
5. Heat alcohol test tube in the boiling water for 15 minutes or until the leaf is white. ***Note:*** You may need to add more alcohol to the test tube if too much alcohol boils away. The alcohol removes the chlorophyll from the leaf.
6. Place the leaf in the petri dish. Add 10-15 drops of iodine directly to the leaf. The leaf should be coated in iodine. Let the iodine soak in for 3 minutes.
7. Dispose of the alcohol in the marked waste container.
8. Rinse the excess iodine off, then fill the petri dish one-third full of water.
9. Spread the leaf out flat.
10. Sketch the leaf indicating the stained areas.
    Note: Iodine stains starch a dark color, so the darkly stained portions of the leaf indicate the presence of starch.
11. Wash and return glassware. Clean lab table.

Name ______________________________ Section ______________

Part III: Data and Discussion1.

1. Paper Chromatography
   a. The chromatography strip must be attached to the lab report of one member of the research group.
   b. The name of each research group member must be listed here:

      (1) ______________________

      (2) ______________________

      (3) ______________________

      (4) ______________________

   c. Label these pigments on the chromatography strip:
      Chlorophyll a – light green color
      Chlorophyll b – dark green color (least soluable)
      Xanthophyll – pale yellow color
      Carotene – bright-yellow color (most soluable).
   d. If your lab report does not have the chromatography strip attached here, draw the strip and label the pigments here.

2. What is the most predominant pigment present in the spinach leaf? Give evidence to support your answer.

3. Why do different pigments move up the chromatography paper at different rates?

Name ______________________________ Section ______________

4. What is the:

   a. most soluble pigment?

   b. least soluble pigment?

5. Pigments Responsible for Photosynthesis

   Diagram the iodine-stained leaf in the space below.

6. Why was the leaf placed in the iodine?

7. Iodine stains starch a dark purple color. Did the leaf test evenly for the presence of starch? Why or why not?

8. Since photosynthesis produces sugar, which can then be stored as starch, what pigment can be associated with photosynthesis? Give evidence to support your answer.

Name ______________________________ Section ______________

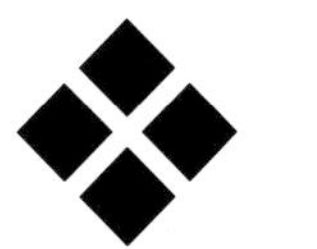

Lab 26

Diversity of Plants and Fungi

Problem

What characteristics are used to identify different plants and fungi?

Objectives

After completing this exercise, the student will be able to:

1. List the classification from domain through species in proper sequence.
2. Recognize the characteristics used to differentiate among the various groups of plants and fungi.
3. List the characteristics of each plant or fungus group.
4. Identify an unknown organism into the proper domain, kingdom, phylum, or class.
5. List characteristics and be able to identify examples of the following groups:

   Kingdom Protista
   - Phylum Chlorophyta

   Kingdom Fungi

   Kingdom Plantae
   - Phylum Bryophyta
   - Phylum Pteridophyta or Polypodiophyta
   - Phylum Gymnospermae or Pinophyta
   - Phylum Angiospermae or Magnoliophyta
     - Class Monocotyledonae or Liliopsida
     - Class Dicotyledonae or Magnoliopsida

Preliminary Information

Current taxonomic classification places an organism in one of three very large groups called **Domains**. The three domains are:

- ❖ Domain Archaea
- ❖ Domain Bacteria
- ❖ Domain Eukarya.

Name ______________________________ Section ______________

The Domains Archaea and Bacteria contain bacteria distinguished by their genetic makeup. It is beyond the scope of this lab to distinguish them. Domain Eukarya is divided into Kingdoms:

- ❖ Domain Eukarya
  - Kingdom Protista
  - Kingdom Fungi
  - Kingdom Plantae
  - Kingdom Animalia.

Part I: Natural System of Plant Classification

The classification system of plants that is most widely accepted today is based on a natural system in which the groupings of plants show how closely related they are. This system is broken down as follows:

Domain	Continent
Kingdom	Country
Phylum	State
Class	Town
Order	Street number
Family	House number
Genus (plural, genera)	Last name
Species	First name

Each category in this hierarchy is a collective unit containing one or more groups from the next lower level in the hierarchy. Thus a phylum is a group of related classes; a class is a group of related orders; an order is a group of related families, etc.

Part II: How to Use a Key

Most keys begin with some obvious characteristic and present you with two choices. In the following key, at step #1, you must decide if the cells are without a nucleus or have a nucleus. If the cells do not have a nucleus, the organism is in the **Domain Bacteria**. Do not go any further since there is no number sending you to another step. If the cells have a nucleus, the organism is in the **Domain Eukarya**, then continue on to step #2. At step #2, you must decide if the organism is single celled or multicelled. A single cell is so small that you need a microscope to see it. Do not confuse a colony made of a chain of similar cells with a multicellular organism which has many specialized parts, such as roots, stems, leaves, digestive tract, or wings. If the organism is multicellular, go to step #3.

Key to Domains and Kingdoms

1a. Cells without a nucleus . . . . **Domain Bacteria**
1b. Cells with a nucleus . . . . **Domain Eukarya**, go to 2

2a. Single cells or colonies of similar cells; usually microscopic . . . . **Kingdom Protista**
2b. Multicelled organisms, tissues or layers of cells; usually macroscopic 3

3a. Green on some part; contains cholorophyll; non-mobile . . . . **Kingdom Plantae**
3b. Brown, tan, or gray; no cholorophyll; feeds on decaying matter . . . . **Kingdome Fungi**
3c. Mobile; eats other organisms . . . . **Kingdom Animalia**

Name ____ Section ________________

Part III: Recognition of Plant and Fungus Characteristics

In the following exercises you will become familiar with some major groups of fungi, algae and plants. You will study the following groups:

1. Fungi
2. Green algae
3. Mosses
4. Ferns
5. Seed plants
   a. Evergreens
   b. Flowering plants
      (1) Monocots
      (2) Dicots

Notice that you will be looking at the subgroups of seed plants in detail. Most of the plants around us belong to this group.

The placement of a plant into one of these groups depends upon the presence or absence of certain characteristics. Before you begin to classify organisms, you must be able to recognize these characteristics.

Examination of Plant and Fungus Characteristics

On the demonstration table you will find materials that will help you to become familiar with certain plant terms.

A. **Chlorophyll**

The presence of the green pigment chlorophyll gives a green color to the organism. Look at a living plant and a fungus on the demonstration table.

What color is the plant? ____________
Does it have chlorophyll? ____________
What color is the fungus? ____________
Does it have chlorophyll? ____________

B. **True Leaves and Stems**

Vascular bundles or **veins** are groups of specialized conductive tubes, which move water, food, minerals, and hormones through a plant. **The presence of vascular tissue is the criterion for determining if a plant has true leaves or stems**. The vascular bundles of a leaf are the veins of the leaf. A single midrib by itself is not evidence of vascular bundles. Observe the leaflike structures of the material on the demonstration table. To establish if true leaves are present examine the bottom of the leaf or hold the leaf up to the light. If you observe veins running throughout the leaf, it is a true leaf. Observe the stringy parts in the stem of a celery stalk. These are vascular bundles or veins.

Examine prepared slides of a cross section through a stem of corn (*Zea*) and mushroom (*Coprinus*), which are available at the demonstration table. Look for the presence of vascular bundles that look like monkey faces. Refer to the Figure 26.1 of cross section of corn stem and locate the vascular bundles.

Vascular tissue is a general name for both **water-conducting tissue (xylem)** and **nutrient-conducting tissue (phloem). The wood of a stem is a concentration of xylem**, which appears as rings in cross section. **The phloem** is just outside the woody xylem in the part commonly called **bark**. Observe the cross section of a tree. Locate the xylem and phloem.

Name____ Section_______________

Which organisms at this demonstration area did not possess vascular tissue? ______________

____

Which organisms at this demonstration possess vasular tissue? ______________________

____

C. Seeds and Spores

Most plants reproduce either by spores or seeds. *Spores* are unicellular; *seeds* are multicellular and are composed of a seed coat, an embryo, and stored food. Spores are produced by fungi, green algae, liverworts, mosses, and ferns. Seeds are produced by the evergreens and flowering plants. At the demonstration table observe the materials labeled seeds and spores.

Which organisms at this demonstration possess vasular tissue? ______________________

____

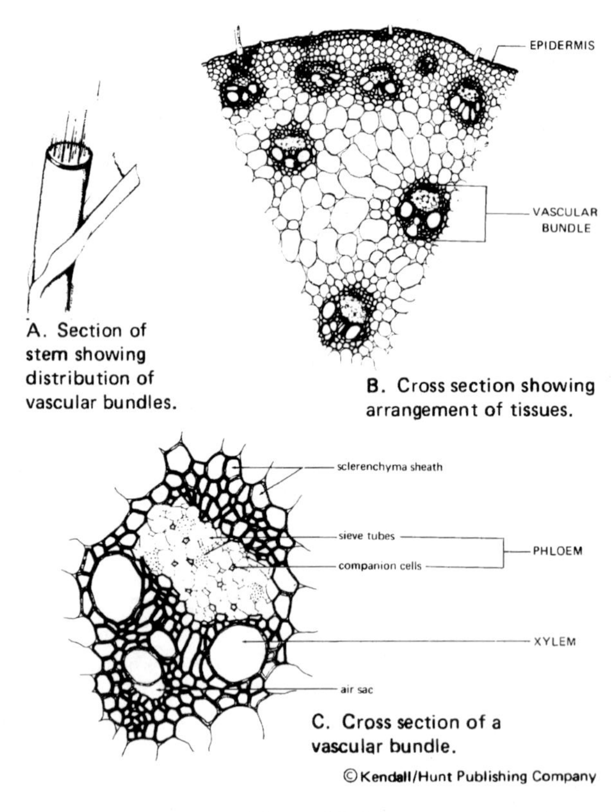

Figure 26.1
Structure of a Monocot (Corn) Stem

Name ______________________ Section ____________

D. **Cotyledons**

Cotyledons are the leaves present on the embryo inside a seed of a flowering plant. There may be one or two cotyledons. While cotyledons are difficult to see, it is possible to easily tell how many cotyledons a flowering plant has by examining other plant parts such as its seeds, flower, or leaves.

All seeds either separate into two halves (dicotyledons) or do not separate easily (monocotyledons). Obtain a soaked bean and corn seed from the supply area. Examine each by taking the seeds apart.

Which is the monocotyledon seed? ______________________

Which is the dicotyledon seed? ______________________

Plants producing monocotyledon seeds are called *monocots* and plants producing dicotyledon seeds are called *dicots*.

E. **Flower Parts**

All covered seeds are borne in modified structures of flowers. Observe a typical flower (Figure 26.2) and notice the circle of brightly colored *petals*, the pollen producing *stamens*, and the vaseshaped *pistil*. **Stamens** are the **male reproductive parts** and the **pistil** is the **female reproductive part** of the flower. A pollen grain from the same flower or a different flower of the same species will fertilize the egg cell within the base of the pistil. The fertilized egg will develop into an embryo. The embryo and its large mass of stored food is called a seed. Thus the seed is completely enveloped by the pistil. It is the **ovary** of the flower, which eventually develops into *fruit*, usually greatly enlarging in the process.

How many flower parts (petals, stamens, or stigmas) are found in the monocot? ______________

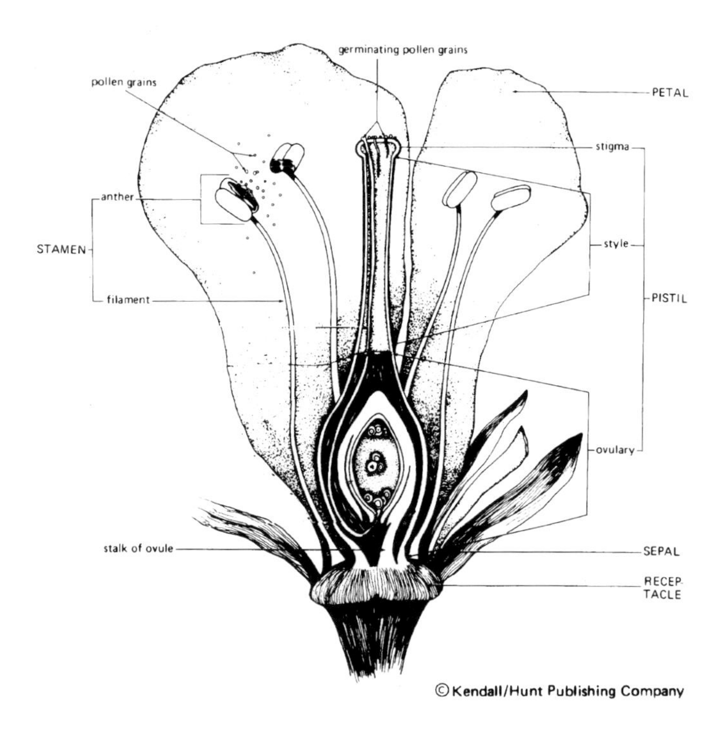

Figure 26.2
Generalized Structure of a Dicot Flower

How many flower parts (petals, stamens, or stigmas) are found in the dicot? ______________

The flower parts (petals, stamens, stigmas) of monocots occur in three, or multiples of three. The flower parts of dicots occur in four or five, or multiples of four or five.

F. **Leaf Venation**

Leaves having veins that lie parallel to one another from the base to the tip of the leaf are *parallel-veined* leaves. Leaves having veins that split off from the main midrib vein so that the venation resembles a net are *net-veined* leaves. Examine the leaves of monocot (spider plant) and dicot (violet).

Does the monocot have parallel-veined leaves or net-veined leaves? ______________

Does a dicot have parallel-veined leaves or net-veined leaves? ______________

Parallel-veined leaves indicate a monocot and net-veined leaves indicate a dicot.

G. **Summary of Plant Characteristics**

A green color indicates the presence of ______________.

True leaves are indicated by the presence of ______________.

Wood is a concentration of ______________.

If a seed, after removing the seed coat, splits into two halves while you are examining it, the plant that produced this seed could be called a ______________.

A (an) ______________ matures and ripens into a fruit containing seeds.

If a flower has three petals, the plant could be called a ______________.

If a plant has net venation and five petals, it could be called a ______________.

Part IV: Classification of Unknown Plants

Using the Taxonomic Key for Plants, Fungi, and Protists, classify the 11 specimens at the demonstration table. Classify each specimen and then check with your instructor to make sure you have classified everything correctly. If any are wrong, go through the key again to locate your error.

This will be valuable practice for the practical quiz that you will have to take. The practical quiz will consist of 10 unknown specimens that you will have to identify to kingdom, phylum, and class *without* the use of the key.

Name ______________________ Section ______________

Taxonomic Key for Plants, Fungi, and Protists

1a. Chlorophyll absent; no true roots, stems, and leaves; often mushroom-shaped. . . . . **Kingdom Fungi**
1b. Chlorophyll present . . . . go to 2

2a. Single-celled organisms or colonies of single cells: aquatic; usually microscopic . . . . **Kingdom Protista, Phylum Chlorophyta**
2b. Multicellular organisms . . . . **Kingdom Plantae:** go to 3

3a. Vascular tissue absent. . . . . **Phylum Bryophyta**
3b. Vascular tissue present . . . . go to 4

4a. Highly lobed leaves (often indented to the midrib) grow from horizontal (often hairy) stems resembling roots; spores on under surface of leaves . . . . **Phylum Pteridophyta**
4b. Needlelike leaves; seeds produced in cones . . . . **Phylum Gymnospermae**
4c. Leaves broad or narrow; seeds produced in flowers. . . . . **Phylum Angiospermae:** go to 5

5a. Parallel-veined leaves; flower parts in threes; seeds with one cotyledon . . . **Class Monocotyledonae**
5b. Net-veined leaves; flower parts in four or fives; seeds with two cotyledons . . . **Class Dicotyledonae**

Table 26.1
Plant Classification Using a Taxonomic Key

Note: Each specimen may not have all the categories.

Specimen Number	Common Plant Name	Kingdom	Phylum	Class
1.	Fern			
2.	Spider Plant			
3.	Pine branch w/cone			
4.	Oak leaf and acorn			
5.	Moss			
6.	Green algae			
7.	Corn leaf w/corn kernels			
8.	Mushroom			
9.	Grass			
10.	Tulip flower			
11.	Violet flower			

Name ______________________ Section ______________

Lab 27

Fungal Diversity

Problem

What are the types of fungi?

Objectives

After completing this lab exercise, the student will be able to:

1. Give examples of unicellular and multicellular fungi.
2. Identify the relationship between fungi and other life forms.
3. Identify the structural components of a fungus.
4. Explain reproduction and metabolism in yeast.
5. Discuss the methods fungi use to obtain nutrition.
6. Explain several methods of fungal reproduction.
7. Name the source of penicillin.
8. Name a symbiotic relationship involving fungi.

Preliminary Information

The Kingdom Fungi contains about 100,000 known species. It is estimated that there may be as many as 300,000 species. They range in size from single-celled yeast to large multicelled organisms. Some of the common fungi are yeast, black bread mold, truffles, morels, bracket fungi, mushrooms, puffballs, *Penicillium*, ringworm, "athlete foot," and lichens. The multicelled fungi are composed of a mass of tubular strands called **hyphae** which branch extensively and fuse together. A mass of fused hyphae is called a **mycelium**.

Some fungi are **saprophytes**, meaning they obtain their energy from nonliving organic matter. Most of the saprophytes (yeast is an exception) secrete digestive enzymes then absorb the pre-digested molecules of life. These fungi are essential recyclers of elements back into the food chain. Examples of recyclers include black bread mold, truffles, morels, bracket fungi, puffballs, mushrooms, and *Penicillium*.

Some fungi are **symbiotic**, living in a mutally beneficial association with another organism, in this case, with algae. The combination of the fungus and algae is called **lichen**. The fungus provides protection and moisture for the algae which produce food for both by photosynthesis.

Other fungi are **parasites**, feeding off a living host. Examples of parasitic fungi are ringworm, "athlete's foot," and *Candida albicans*, which causes vaginal yeast infection and thrush (fungal infection in the mouth).

Name ______________________ Section ____________

Microscope Reminder

Compound Microscope:

- ❖ Typically one ocular at 10X magnification
- ❖ Revolving nosepiece with three objective lens:
  - Scanner, 4X (shortest lens)
  - Low power, 10X
  - High power, 43X (longest lens)
- ❖ **Total magnification** = objective X ocular, such as total magnification using high power = 43 X 10 or 430 total magnifications.

Part I: Yeast

Yeasts are single-celled fungi with the unusual ability to dry out and stop their metabolism. When rehydrated, they begin cell respiration again. Yeasts also have the unusual ability to switch between aerobic respiration (used when oxygen is present) and anaerobic respiration (used when no oxygen is present).

Aerobic respiration produces water, carbon dioxide, 36 ATP molecules, and heat:

$$C_6H_{12}O_6 + 6\,O_2 \rightarrow 6\,H_2O + 6\,CO_2 + 36\text{ATP (chemical energy)} + \text{heat energy.}$$

Anaerobic respiration in yeast produces ethyl alcohol, carbon dioxide, 2 ATP, and heat:

$$C_6H_{12}O_6 \rightarrow 2\,C_2H_5OH + 2\,CO_2 + 2\text{ ATP (chemical energy)} + \text{heat energy.}$$

The ATPs are used to power all cell activities.

Yeasts reproduce asexually by **budding**. A new small yeast cell (a bud) begins to grow from the side of the larger parent yeast cell. It enlarges and breaks off from the parent. It is common for a second bud to form on the first bud before it breaks off from the original parent.

1. Obtain a plain microscope slide and cover slip. Wash and dry the slide and cover slip.
2. Stir the yeast solution to eliminate bubbles on the top.
3. Add a drop of yeast solution to the microscope slide using a pipette.
4. Add a cover slip.
5. Observe using high power (430X) of the compound microscope.
   Note: The yeasts are carrying on anaerobic respiration converting sugar to ethyl alcohol and carbon dioxide. Each carbon dioxide gas bubble appears as a sphere with a heavy black edge.

Name ______________________________ Section ______________

6. Diagram and label:
   - **yeast**
   - **bud**

Yeast

Part II: Black Bread Mold

Black bread mold is commonly found growing on old bread, usually without preservatives, that has been kept in a warm, damp environment. Black bread mold has both asexual and sexual reproductive modes. In asexual reproduction, the fungus reproduces using spores. The spores become black as they mature, giving the mold the "black" portion of its name. When the spores are released and blown to a new location, they can each grow to form a new mycelium.

1. Examine a model of black bread mold.
2. Examine a prepared microscope slide of black bread mold.
3. Diagram and label:
   - the thread-like **hyphae** combining to form the mycelium,
   - the **sporangiophore** (stalk) with the **sporangium** (ball) on the tip,
   - black **spores** formed in each sporangium.

Black Bread Mold

Model **Prepared Slide**

Name ______________________________ Section ______________

Part III: Penicillium

Penicillium is a mold commonly found growing on decaying fruit, such as oranges. This mold produces penicillin, the well-known antibiotic. The antibiotic characteristic of this mold was discovered in 1929 by Sir Alexander Fleming, an English researcher.

1. Examine and diagram a model of *Penicillium.*
2. Obtain a prepared microscope slide of Penicillium: conidia, w.m.
3. Using high power (430X) of the compound microscope, examine the edges of the mass.
4. Diagram and label:
   - ❖ the rows of finger-like hyphae called **conidia**,
   - ❖ the **spores** contained in the conidia.

Penicillium

Model **Prepared Slide**

Name ______________________________ Section ______________

Part IV: Coprinus

Coprinus is the common mushroom. Most of the fungus grows in the soil. When conditions are correct, the fungus grows a reproductive structure above ground which produces spores; this is the mushroom. The mushroom has a large central stem supporting the dome-shaped cap. The spores are produced on club-shaped structures called **basidia** which grow from the gills on the underside of the cap. Each basidium produces four **basidiospores**. These spores are blown by the wind to a new location where a new fungus develops from each spore.

1. Obtain a fresh mushroom for your lab table. Observe its underside.
2. Diagram and label:
   - ❖ the **stalk** in the center,
   - ❖ the outer **cap** with its dark brown **gills**.
3. Obtain a prepared microscope slide of Coprinus, x.s.
4. Using scanner power (40X) of the compound microscope, examine the slide.
5. Diagram and label:
   - ❖ the **stalk** in the center,
   - ❖ the outer **cap** with its **gills** radiating toward the stalk, like the spokes of a wheel.
   - ❖ Notice the cells in the stalk are all of the same design. There are no specialized cells for water conduction or food movement that are found in the stems of higher plants. Fungi have no vascular tissue for transferring water or nutrients around the fungal body. Water and nutrients move through the cells of the mycelium so the movement is slower than in vascular plants.
6. Using high power (430X) of a compound microscope, examine the edge of a gill.
7. Diagram and label:
   - ❖ the oval **basidiospores** (dispersed by the wind),
   - ❖ the club-like base, the **basidium**, from which the basidiospores emerge.

Coprinus

Fresh

Prepared Slide

Scanner power of x.s.

High power of basidiospores

Name ______________________________ Section ______________

Part V: Lichen

Lichen is a combination of fungus and algae living together as one unit. They live in a mutualistic relationship where both organisms benefit. The fungus provides chemical substances and moisture for the algae. The algae, living between the hyphae, supply food, through photosynthesis, for themselves and the fungus.

1. Observe the demonstrations of lichens.
2. Diagram and label the lichen types as:
   - ❖ **Crustose** (forming a fine crusty appearance),
   - ❖ **Foliose** (forming a leaf-like mat),
   - ❖ **Fruiticose** (having a more erect shrub-like or plant-like appearance).
3. Obtain a prepared slide of *Physcia*.
4. Using low power (100X) of a compound microscope, scan the middle part of the lichen for a clear view of the hyphae and the algae. Switch to high power (430X). Look for small round algae and the tube-like fungal hyphae.
5. Diagram and label:
   - ❖ **fungal hyphae**,
   - ❖ **algae**, the round or oval single cells between the hyphae.

Lichens

Types | **Microscopic Structure**

Name ______________________________________ Section ________________

Part VI: Larger Fungi

1. Observe the fungi on display at the demonstration area.
2. Diagram and identify two different types of fungi.

   Fungus Name: ____________________

   Fungus Name: ____________________

3. Which have you diagrammed?
   - The mycelium.
   - The reproductive structures.

Name ____ Section ______________

Part VII: Discussion

1. Indicate the total magnifications for the compound microscope using:
   - Scanner = ______________
   - Low power = ______________
   - High power = ______________
2. What is the gas in the bubbles produced by the yeast?
3. How does yeast reproduce?
4. Multicellular fungi are composed of thread-like structures called ______________.
5. The fungal mass (body) formed by the thread-like structures is the ______________.
6. Multicellular fungi reproduce and disperse themselves using ______________.
7. Since black bread mold grows on white bread, why is it called black bread mold?
8. What is penicillin and what produces it?
9. What type of organism is lichen?

Name ______________________________ Section ______________

Lab 28

Animal Classification

Problem

How can organisms be grouped into systematic categories according to their physical characteristics?

Objectives

After completing this laboratory exercise, the student will be able to:

1. Recognize the diversity between major animal groups and within major animal groups.
2. List the taxonomic sequence from domain through species.
3. Determine if an organism has radial, bilateral, or no symmetry.
4. Determine if an organism has an endoskeleton, an exoskeleton, or no skeleton.
5. Recognize complete or partial segmentation.
6. Identify an animal into phylum or class using a taxonomic key.

Preliminary Information

One of the first persons to group organisms was Aristotle. Aristotle grouped animals according to where they lived. This method was confusing and did not show relationships. For example, a whale, a shark, and an octopus would be classified together. But it was not until the 1700s that *taxonomy* or the science of classification really developed. During this time, a Swedish botanist, Carolus Linneaus, developed the system of *binomial nomenclature*, identifying an organism by genus and species. The advantage of this system is that every living thing is given only one name and no two kinds of organisms can have the same name. This eliminates the confusion of common names. For instance, the bird Americans refer to as a robin is called a thrush in England, and in Europe it has a different name in every country. However, if you use the name *Turdus migratorius*, it can mean only one type of bird, no matter if you are American, English, Russian, or Japanese.

Part I: Taxonomic Rank

Placing organisms in categories can be like sorting mail.

Domain	Continent
Kingdom	Country
Phylum	State
Class	Town
Order	Street number
Family	House number
Genus	Last name
Species	First name

Name ______________________________ Section ______________

	Human	**Wolf**	**Herring Gull**	**Red Oak**
Domain	Eukarya	Eukarya	Eukarya	Eukarya
Kingdom	Animalia	Animalia	Animalia	Plantae
Phylum	Chordata	Chordata	Chordata	Magnoliophyta
Class	Mammalia	Mammalia	Aves	Dicotyledonae
Order	Primates	Carnivora	Charadriiformes	Fagales
Family	Hominidae	Canidae	Laridae	Fagaceae
Genus	*Homo*	*Canis*	*Larus*	*Quercus*
Species	*sapiens*	*lupus*	*argentatus*	*ruba*

The importance of this type of a classification scheme is that it shows the relationship between organisms. So if two organisms belong to the same family, but different genera, they are more closely related than two organisms that belong to the same order, but different families.

1. Of the other three organisms classified in the above table, which is most closely related to humans?Why? ______________________________
2. Of the other three organisms classified in the above table, which is least related to humans? Why?

____

Part II: Recognition of Characteristics

Before you begin using a more specific key, you should be able to recognize certain characteristics. Standard characteristics are generally used in classification. Identify organisms with the following characteristics. **Use the demonstration specimens and animal set specimens.** If you do not recognize the organism, match its identification number with its name by using the list at the end of this lab.

Single-celled organisms or colonies: Single-celled organisms exist by themselves. Colonies are groups of single-celled organisms attached and living together but physiologically independent. Each cell in a colony looks like every other cell. The cells are not specialized into different functions. Colonies often form thread-like strands. Go to the demonstration table, examine the specimens and name two that are single-celled or colonies. ______________________________

Radial symmetry: Many lines could be drawn top to bottom through the middle of an animal and two identical halves would result. These organisms tend to be circular or radiate out from a central point. Refer to Figure 28.1. Name three organisms with radial symmetry. ______________________________

Bilateral symmetry: Only one line can be drawn top to bottom through the middle of an organism that will result in two identical external halves. Refer to Figure 28.1. For instance, in humans the only way you can get two identical halves is by passing a line down the middle through the nose, mouth, etc. Name three other organisms with bilateral symmetry. ______________________________

Name ______________________________ Section ______________

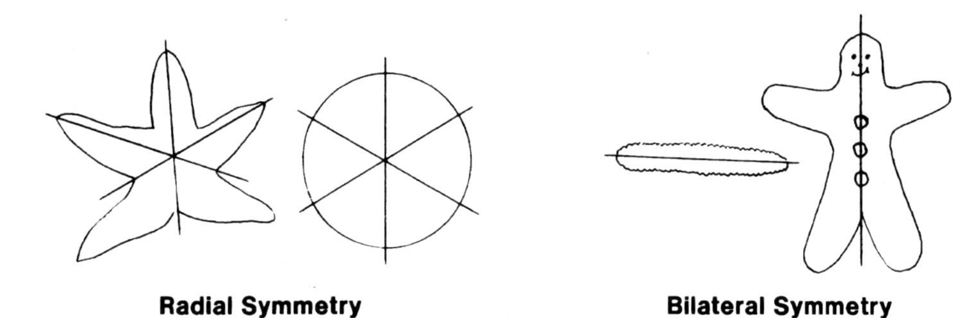

Radial Symmetry

Bilateral Symmetry

Figure 28.1
Symmetry

No symmetry: No particular shape. Name one organism that lacks symmetry.

Segmentation: Segments are repeating units that are similar. With complete segmentation the units are obvious, like connected railroad cars. Incomplete segmentation is not so obvious. The segmentation need not be external, it may be part of the skeleton such as the backbones in a human.

Name three organisms that show some degree of segmentation, internal or external.

____

No Segmentation: Name three organisms that lack any visible segmentation. ______________

____

Exoskeleton: Hard outer covering, these animals will usually go crunch when stepped on. Name three.

____

Endoskeleton: Internal bony structure. Examine demonstrations of animal skeletons. Name three animals that have an endoskeleton. ______________

____

Hydrostatic skeleton: Rigidity is provided when muscles squeeze fluids in the body. There is no specific skeletal structure. Animals with hydrostatic skeletons usually have segments to increase hydraulic pressure. Name two animals that have a hydrostatic skeleton. ______________

____

Name ______________________________ Section ______________

Part IV: Classification of the Animal Kingdom

Examine all of the specimens in the set. Notice they are all animals. So now you are ready to proceed to the next level of classification, the phylum.

The first two phyla groups have been suggested for you. Separate all the jars into separate groups that fit this description. Write the number of the jar after the appropriate description. These two phyla will consist of *more than half* of the jars. After you have completed these two groups to your satisfaction, have the *instructor check* before proceeding.

Now work with the remaining jars. Place those that look alike together. Devise a description of the group and record it as phylum number 3. Again record the number of the jars. These remaining groups will not be as large as the first two. Make sure all organisms are classified. It is important that you describe how the organism looks. For example, "long, slender, round, with segments" is fine. But it is not permissible to state "earthworm" as it does not tell you how an earthworm looks.

Do not use your textbook or any other reference.

You will not necessarily need to form 10 phylum groups. The number of groups can vary depending on the characteristics that you use to identify the phyla.

Phylum Level

1. Exoskeleton and jointed appendages

2. Endoskeleton

3.

4.

5.

6.

7.

8.

9.

10.

When you complete this portion, have your *instructor check* your groups before you proceed. The first two phyla groups have been selected for you to subdivide into classes. Again *describe* but do not name the group. It is not necessary to have six classes in each phylum. The number of classes depends on the characteristics you use to set up the classes.

Name ______________________________ Section ______________

Classes for exoskeletons and jointed legs.

1.
2.
3.
4.
5.
6.

Classes for endoskeletons.

1.
2.
3.
4.
5.
6.

Part V: Use of a Key

The type of key that will be utilized is referred to as a dichotomous or branching key, which means at every level of the key you always have at least two choices. Read the description and decide which one best fits the specimen you are keying. To the right of each description you are referred by number to the next part of the key or the classification name is given.

Classify each specimen. Check with your instructor to make sure you have classified everything correctly. If any are wrong go through the key again to locate your error. This will be valuable practice for the practical quiz that you will be taking. The *practical quiz* will consist of 20 unknown specimens that you will have to identify to kingdom, phylum, and class *without the use of the key*.

Name ______________________ Section ______________

A Key for the Kingdom Animalia

1a. Body lacks symmetry; body porous .... **Phylum Porifera**
1b. Body with radial symmetry 2
1c. Body with bilateral symmetry 3

2a. Body soft; disc, tube or saclike with radiating tentacles .... **Phylum Cnidaria**
2b. Body hard, spiny or leathery; body parts may be in fives .... **Phylum Echinodermata**

3a. Endoskeletons with backbones or cartilage; appendages when present of two pairs .... **Phylum Chordata (Subphylum Vertebrata)** (go to Chordata Key)
3b. Exoskeletons; appendages, when present, of three or more pairs 4
3c. No skeleton or legs 4

4a. Jointed legs; exoskeletons .... **Phylum Arthropoda** (go to Arthropoda Key)
4b. Without jointed legs; no-skeleton, but shell may be present 5

5a. Body many times longer than wide, but no shell 6
5b. Body in a shell;
protrusible structure used for locomotion .... **Phylum Mollusca**

6a. Body flat; with or without segmentation .... **Phylum Platyhelminthes**
6b. Body round when viewed from end 7

7a. Body with segmentation .... **Phylum Annelida**
7b. Body without segmentation .... **Phylum Nematoda**

A Key of the Phylum Arthropoda

1a. Paired antennae present 2
1b. Paired antennae absent; four pairs of walking legs .... **Class Arachnida**
1c. Paired antennae absent; five pairs of walking legs with claws .... **Class Merostomata**

2a. One pair of antennae 3
2b. Two pair of antennae; one long, one short .... **Subphylum Crustacea**

3a. Three pair of walking legs .... **Class Insecta**
3b. More than three pair of walking legs .... **Subphylum Myriapoda**

A Key of the Phylum Chordata, Subphylum Vertebrata

1a. Scales and fins .... **Class Osteichthyes**
1b. No fins, legs usually present 2

2a. Body covering soft and moist; if feet, without claws .... **Class Amphibia**
2b. Body covering dry and scaly; if feet, with claws .... **Class Reptilia**
2c. Body covering feathers .... **Class Aves**
2d. Body covering hair .... **Class Mammalia**

Name ______________________________ Section ______________

Common Name	Phylum	Subphylum or Class (if appropriate)
1. Shrimp		
2. Sand dollar		
3. Sea urchin		
4. Planaria		
5. Roundworm		
6. Crayfish		
7. Moon jelly		
8. Mouse		
9. Fish		
10. Beetle		
11. Cockroach		
12. Crab		
13. Sea anemone		
14. Razor clam		
15. Hydra		
16. Bat		
17. Spider		
18. Bird		
19. Grasshopper		
20. Salamander		
21. Moon jelly		
22. Lizard		

Name ______________________________ Section ______________

23. Purple sea urchin ______________ ______________

24. Oyster ______________ ______________

25. Earthworm ______________ ______________

26. Chiton ______________ ______________

27. Sponge ______________ ______________

28. Scorpion ______________ ______________

29. Leech ______________ ______________

30. Long neck clam ______________ ______________

31. Liver fluke ______________ ______________

32. Turtle ______________ ______________

33. Portugese man-of-war ______________ ______________

34. Frog ______________ ______________

35. Sea star ______________ ______________

36. Roundworm ______________ ______________

37. Scallop ______________ ______________

38. Brittle star ______________ ______________

39. Millipede ______________ ______________

40. Snake ______________ ______________

41. Centipede ______________ ______________

42. Snail ______________ ______________

43. Horseshoe crab ______________ ______________

44. Tapeworm ______________ ______________

45. Sandworm ______________ ______________

Name ______________________________ Section ______________

Lab 29

Electrocardiogram and Blood Pressure

Problem

How does heart beat, blood pressure, and respiration relate in maintaining homeostasis in humans?

Objectives

After completing this lab, the student will be able to:

1. Identify the three main parts of a normal EKG (electrocardiogram).
2. Observe effects of muscular contractions on an EKG.
3. Measure vital capacity using a spirometer.
4. Measure pulse.
5. Determine human blood pressure through the use of a sphygmomanometer and stethoscope.
6. Construct tables and graphs from data obtained from blood pressure and pulse measurements.

Preliminary Information

When the heart contracts, blood is pumped to the lungs where carbon dioxide is given off and oxygen is picked up. The carbon dioxide is expelled and new air is drawn into the lungs by the respiratory movements of the diaphragm and muscles, which move the ribs.

When the hearts contracts, blood is pumped under pressure to all parts of the body. Blood pressure can be used as an important diagnostic tool and as an indicator of general physical health.

When the heart contracts, electric currents spread into and over the heart. Some of these electric currents spread over the surface of the body and can be recorded from electrodes placed on the body.

The electrocardiograph is an instrument that amplifies and records the voltages produced by the heart as it contracts and relaxes. The electrocardiogram (EKG) is the record that is produced by the electrocardiograph.

Part I: Electrocardiogram—EKG (ECG)

The typical EKG records P, QRS, and T waves. The **P wave** represents contraction (excitation) of the atria; **QRS wave** indicates contraction (excitation or depolarization) of the ventricles; the **T wave** indicates relaxation (repolarization) of the ventricles.

By analyzing these wave forms, certain deductions can be made which offer enormous information about the heart.

Name ____ Section ______________

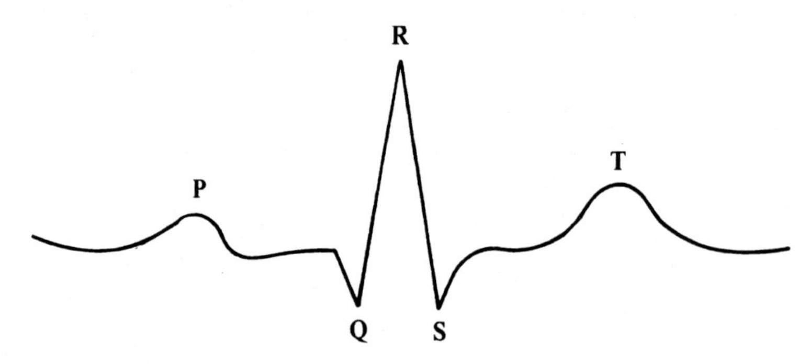

Figure 29.1
Configuration of the Normal EKG

Procedure for EKG

1. Have the subject sit in a comfortable position.
2. Using alcohol and cotton or an alcohol pad, clean the skin at the sites of electrode attachment (described below), because natural oils of the skin act as insulation.
3. Open one of the packages of electrolyte pads. These pads are damp with a solution allowing good electrical conduction between skin and electrode.
4. Place one pad on each electrode indicated below, then **attach the electrode over a bone** near the site indicated on the back of the electrode:
   RA—attach to bone just above wrist of right arm
   LA—attach to bone just above wrist of left arm
   RL—attach to bone on front of right leg
   LL—attach to bone on front of left leg
   C—sit on the brown lead.
5. Electrocardiograph adjustments:
   a. Turn main power switch on.
   b. Set the sensitivity switch on 1.
   c. Set the lead control switch on 2.
   d. To begin taking the EKG, move the center control lever to Run 25.
   NOTE: Abnormal readings indicate an electrode may not be firmly attached.
6. Normal EKG: Have the subject sit quietly to obtain the normal EKG pattern. Run the electrocardiograph for about 20 seconds.
7. Effect of muscle contraction: While keeping other muscles relaxed, have the subject clench one fist. Note: The heart is not affected but the EKG strip is distored since it is recording muscle activity.
8. On the electrocardiogram, label:
   a. **Your name**
   b. **atria contracting**
   c. **ventricle contracting**
   d. **ventricle relaxing**
   e. **possible abnormalities** with a circle
   f. **fist contraction.**
9. Attach the EKG strip to this page below the diagram of the normal EKG.

Name ______________________________ Section ______________

Part II: Pulmonary Function

Air is drawn into the lungs by a combination of the action of muscles between the ribs and the diaphragm muscles across the abdomen. The physical dimensions of your lungs vary according to heredity, sex, age, smoking, exercise, and other factors.

The air in your lungs can be defined in the following way:

- ❖ Tidal Volume: amount of air inspired or expired with each normal breath.
- ❖ Expiratory Reserve Volume: air that can still be expired by forceful expiration after the end of normal tidal expiration.
- ❖ Vital Capacity: the maximum volume of air that can be exhaled after taking a deep breath.

1. Put a clean disposable cardboard mouthpiece on the spirometer.
2. Turn the face of the spirometer until the red zero mark lines up with the needle.
3. Air exhaled through the spirometer is recorded in cubic centimeters (cc) by the movement of the needle.
4. To measure vital capacity:
   a. Inhale as much as you can,
   b. hold your nose closed,
   c. exhale as much as you can through the spirometer.
5. Compare your results with the "Percentage of Vital Capacity Chart."
   Note: Reading the "Percent of Vital Capacity Chart":
   a. Find the table for your gender,
   b. Find your "Standing height" in inches (ins) line on the left side,
   c. Find your "Cubic centimeters" column across the top,
   d. Where your line and column intersect, the "Percent of Vital Capacity" is given.
6. Record the data for all members of your group in the table.

Table 29.1
Spirometer Data

Subject Name	Sex	Height	Vital Capacity	Percent Vital Capacity[1]
1				
2				
3				
4				

(1) Use Table 29.2

Name ____________________ Section ____________________

Table 29.2
Percentage of Vital Capacity Chart

PERCENTAGE OF VITAL CAPACITY — MEN — (CALCULATED FROM STANDING HEIGHT) IN CUBIC CENTIMETERS

Standing height in cms.	ins.	700	800	900	1000	1100	1200	1300	1400	1500	1600	1700	1800	1900	2000	2100	2200	2300	2400	2500	2600	2700	2800	2900	3000	3100	3200	3300	3400	3500	3600	3700	3800	3900	4000	4100	4200	4300	4400	4500	4600	4700	4800	4900	5000	5100	5200	5300	5400	5500	5600	5700	5800
144.8	57	19	22	25	28	30	33	36	39	41	44	47	50	52	55	58	61	64	66	69	72	75	77	80	83	86	88	91	94	97	99	102	105	108	110	113	116	119	121														
147.3	58	19	22	24	27	30	33	35	38	41	43	46	49	52	54	57	60	63	65	68	71	73	76	79	82	84	87	90	92	95	98	100	103	106	109	111	114	117	120	122													
149.8	59		21	24	27	30	32	35	37	40	43	45	48	51	53	56	59	61	64	67	69	72	75	77	80	83	85	88	91	93	96	99	101	104	107	109	112	115	117	120	123												
152.4	60		21	24	26	29	32	34	37	39	42	45	47	50	53	55	58	60	63	66	68	71	74	76	79	81	84	87	89	92	95	97	100	102	105	108	110	113	115	118	121	123											
154.9	61		21	23	26	28	31	34	36	39	41	44	47	49	52	54	57	59	62	65	67	70	72	75	78	80	83	85	88	90	93	96	98	101	103	106	108	111	114	116	119	121	124										
157.4	62		20	23	25	28	30	33	36	38	41	43	46	48	51	53	56	58	61	64	66	69	71	74	76	79	81	84	86	89	91	94	97	99	102	104	107	109	112	114	117	119	122										
160.0	63		20	23	25	28	30	33	35	38	40	43	45	48	50	53	55	58	60	63	65	68	70	73	75	78	80	83	85	88	90	93	95	98	100	103	105	108	110	113	115	118	120	123									
162.5	64		20	22	25	27	30	32	35	37	39	42	44	47	49	52	54	57	59	62	64	67	69	71	74	76	79	81	84	86	89	91	94	96	99	101	103	106	108	111	113	116	118	120	123								
165.1	65			22	24	27	29	32	34	36	39	41	44	46	48	51	53	56	58	61	63	66	68	70	73	75	78	80	82	85	87	90	92	95	97	99	102	104	107	109	111	114	116	119	121	123							
167.6	66			21	24	26	29	31	33	36	38	41	43	45	48	50	52	55	57	60	62	64	67	69	71	74	76	79	81	83	86	88	91	93	95	98	100	102	105	107	110	112	114	117	119	121	124						
170.2	67			21	24	26	28	31	33	35	38	40	42	45	47	49	52	54	56	59	61	63	66	68	71	73	75	78	80	82	85	87	89	92	94	97	99	101	104	106	108	111	113	115	118	120	122						
172.7	68			21	23	26	28	30	32	35	37	39	42	44	46	49	51	53	56	58	60	63	65	67	70	72	74	76	79	81	83	86	88	90	93	95	97	100	102	104	107	109	111	114	116	118	120	123					
175.5	69			21	23	25	27	30	32	34	37	39	41	43	46	48	50	52	55	57	59	62	64	66	68	71	73	75	78	80	82	84	87	89	91	94	96	98	100	103	105	107	109	112	114	116	119	121	123				
177.8	70			20	22	25	27	29	31	34	36	38	40	43	45	47	49	52	54	56	58	61	63	65	67	70	72	74	76	79	81	83	86	88	90	92	94	97	99	101	103	106	108	110	112	115	117	119	121	124			
180.4	71			20	22	24	27	29	31	33	35	38	40	42	44	47	49	51	53	55	58	60	62	64	67	69	71	73	75	78	80	82	84	86	89	91	93	95	98	100	102	104	106	109	111	113	115	117	120	122	124		
182.9	72			20	22	24	26	28	31	33	35	37	39	42	44	46	48	50	53	55	57	59	61	63	66	68	70	72	74	77	79	81	83	85	88	90	92	94	96	98	101	103	105	107	109	111	114	116	118	120	122	125	
185.4	73			19	22	24	26	28	30	32	35	37	39	41	43	45	47	50	52	54	56	58	60	63	65	67	69	71	73	76	78	80	82	84	86	89	91	93	95	97	99	101	103	106	108	110	112	114	117	119	121	123	
188.0	74			19	21	23	26	28	30	32	34	36	38	40	43	45	47	49	51	53	55	57	60	62	64	66	68	70	72	75	77	79	81	83	85	87	89	92	94	96	98	100	102	104	106	108	111	113	116	117	119	121	123
190.5	75				21	23	25	27	29	32	34	36	38	40	42	44	46	48	50	53	55	57	59	61	63	65	67	69	71	73	76	78	80	82	84	86	88	90	92	95	97	99	101	103	105	107	109	111	113	116	118	120	122
193.0	76				21	23	25	27	29	31	33	35	37	39	42	44	46	48	50	52	54	56	58	60	62	64	66	68	70	72	75	77	79	81	83	85	87	89	91	93	95	97	99	101	104	106	108	110	112	114	116	118	120
		42.7	48.8	54.9	61.0	62.1	73.2	79.3	85.4	91.5	97.6	103.7	109.8	115.9	122.0	128.1	134.2	40.3	146.4	152.5	158.6	164.7	170.8	176.9	183.0	189.1	195.2	201.3	207.4	213.5	219.6	225.7	231.8	237.9	244.0	250.1	256.2	262.3	268.4	274.5	280.6	286.7	292.8	298.9	305.0	311.1	317.2	323.3	329.4	335.5	341.6	347.7	353.8

PERCENTAGE OF VITAL CAPACITY IN CUBIC INCHES

PERCENTAGE OF VITAL CAPACITY — WOMEN — (CALCULATED FROM STANDING HEIGHT) IN CUBIC CENTIMETERS

Standing Height in cm.	Standing Height in inches	600	700	800	900	1000	1100	1200	1300	1400	1500	1600	1700	1800	1900	2000	2100	2200	2300	2400	2500	2600	2700	2800	2900	3000	3100	3200	3300	3400	3500	3600	3700	3800	3900	4000	4100	4200	4300	4400	4500
139.7	55	21	23	29	32	36	39	43	47	50	54	57	61	64	68	71	75	79	82	86	89	93	96	100	103	107	111	114	118	121											
142.2	56	21	23	28	32	33	39	42	46	49	53	56	60	63	67	70	74	77	81	84	88	91	95	98	102	105	109	112	116	119	123										
144.8	57	21	24	28	31	35	38	41	45	48	52	55	59	62	66	69	72	76	79	83	86	90	93	97	100	103	107	110	114	117	121										
147.3	58	20	24	27	31	34	38	41	44	47	51	54	58	61	64	68	71	75	78	81	85	88	91	95	98	102	105	108	112	115	118	122									
149.8	59	20	23	27	30	33	37	40	43	47	50	53	57	60	63	67	70	73	77	80	83	87	90	93	97	100	103	107	110	113	117	120	120								
152.4	60	20	23	26	30	33	36	39	43	46	49	52	56	59	62	66	69	72	75	79	82	85	88	92	95	98	102	105	108	112	115	118	121								
154.9	61		23	26	29	32	36	39	42	45	48	52	55	58	61	65	68	71	74	77	81	84	87	90	94	97	100	103	106	110	113	116	119	122							
157.4	62		22	25	29	32	35	36	41	44	48	51	54	57	60	63	67	70	73	76	79	83	86	89	92	95	98	102	105	108	111	114	118	121							
160.0	63		22	25	28	31	35	38	41	44	47	50	53	56	59	62	66	69	72	75	78	81	84	88	91	94	97	100	103	106	109	112	116	119	122						
162.5	64		22	25	28	31	34	37	40	43	46	49	52	55	58	62	65	68	71	74	77	80	83	86	89	92	95	98	101	105	108	111	114	117	120						
165.1	65		21	24	27	30	33	36	39	42	46	49	52	55	58	61	64	67	70	73	76	79	82	85	88	91	94	97	100	103	106	109	112	115	118	121					
167.6	66		21	24	27	30	33	36	39	42	45	48	51	54	57	60	63	66	69	72	75	78	81	84	87	90	93	96	98	101	104	107	110	113	116	119	122				
170.2	67		21	24	27	29	32	35	38	41	44	47	50	53	56	59	62	65	68	71	74	77	79	82	85	88	91	94	97	100	103	106	109	112	115	118	120	123			
172.7	68		20	23	26	29	32	35	38	41	44	46	49	52	55	58	61	64	67	69	72	75	78	81	84	87	90	93	96	98	101	104	107	110	113	116	119	121			
175.3	69		20	23	26	29	31	34	37	40	43	46	49	51	54	57	60	63	66	68	71	74	77	80	83	85	88	91	94	97	100	102	105	108	111	114	117	120	123		
177.8	70		20	22	25	28	31	34	37	39	42	45	48	51	53	56	59	62	65	67	70	73	76	79	82	84	87	90	93	96	98	101	104	107	110	112	115	118	121		
180.4	71			22	25	28	31	33	36	39	42	44	47	50	53	55	58	61	64	66	69	72	75	78	80	83	86	89	91	94	97	100	102	105	108	111	114	116	119	122	
182.9	72			22	25	27	30	33	36	38	41	44	47	49	52	55	57	60	63	66	68	71	74	77	79	82	85	88	90	93	96	98	101	104	107	109	112	115	118	120	
185.4	73			22	24	27	30	32	35	38	40	43	46	49	51	54	57	59	62	65	67	70	73	76	78	81	84	86	89	92	94	97	100	102	105	108	111	113	116	119	121
188.0	74			21	24	27	29	32	35	37	40	43	45	48	51	53	56	59	61	64	67	69	72	75	77	80	83	85	88	90	93	96	99	101	104	106	109	112	114	117	120
		36.6	42.7	48.8	54.9	61.0	62.1	73.2	79.3	85.4	91.5	97.6	103.7	109.8	115.9	122.0	128.1	134.2	140.3	146.4	152.5	158.6	164.7	170.8	176.9	183.0	189.1	195.2	201.3	207.4	213.5	219.6	225.7	231.8	237.9	244.0	250.1	256.2	262.3	268.4	274.5

PERCENTAGE OF VITAL CAPACITY IN CUBIC INCHES

Name____ Section______________

Part III: Pulse Rate Measurement

Normal resting pulse rate is between 60 and 100 beats per minute in an adult. Some athletes have pulse rates as low as 40 beats per minute.

1. Working in groups of two, determine each others pulse rate by placing two fingers (not a thumb) over the main artery in the wrist, located just below the upper bone in the forearm.
   Once you have located the artery and felt the pulse, count the pulse rate for 60 seconds. Record your pulse rate in the "Pulse and Blood Pressure" table. Record all group members.
   Optional:
2. Working in either small groups or as a class, measure the pulse rate and blood pressure of a smoker both before and after smoking a cigarette. Step out of the building to smoke. Record your findings on the data sheet.

Optional

Table 29.3
Smoker's Pulse and Blood Pressure

	Pulse	**Blood Pressure**
Normal		
After Smoking		

Name ______________________________ Section ______________

Part IV: Human Blood Pressure Determinations by Sphygmomanometers

A. Apparatus

A sphygmomanometer consists of (1) a compression bag surrounded by an unyielding cuff for application of an extra-arterial pressure, (2) an inflating bulb, pump, or other device by which pressure is created in the system, and (3) a manometer by which the applied pressure is read, (4) a variable, controllable exhaust by which the system can be deflated either gradually or rapidly.

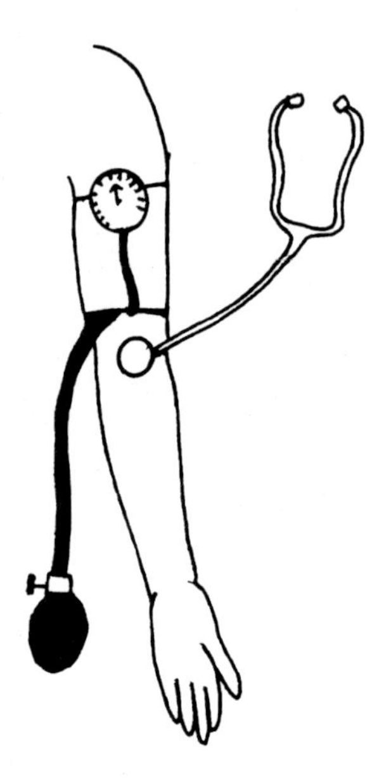

Figure 29.2
Position of Sphygmomanometer and Stethoscope

B. Technique

1. Read and understand each point before beginning. The sphygmomanometer should be inflated only about 45 seconds or less.
2. The patient. The subject should be comfortably seated. The arm should be bared, slightly flexed, and perfectly relaxed. In the sitting position the forearm should be supported at heart level. Keep the arm off the table to avoid hearing artifact noises. The deflated bag and cuff should be applied evenly and snugly around the arm with the lower edge about 3 cm above the antecubital space (inside of elbow) with the rubber tubing attached to the pressure gauge in line with the antecubital space. Anyone who has high blood pressure should not participate in the experiment.
3. Determination of systolic pressure. Clean stethoscope ear pieces with alcohol and cotton or alcohol pad to prevent possible ear infection. Put the stethoscope on and allow a minute to familiarize yourself with the sounds heard through the stethoscope by listening to your own heart.
   In taking a blood pressure, the stethoscope receiver should be applied snugly over the artery in the antecubital space, free from contact with the cuff. See Figure 29.2. The pressure in the sphygmomanometer should then be raised rapidly to about 100 by tightening the screw and pumping up the cuff. Allow a minute to familiarize yourself with the sound of the heartbeat through the artery in the elbow. Now rapidly pump up to 160. Decrease *slowly* by slightly loosening the screw until a sound is heard with each heartbeat. Note the reading on the pressure gauge and record in space below as systolic pressure.

Name ______________________________ Section ______________

4. Determination of diastolic pressure. With continued deflation of the system below systolic pressure at a rate of 2 to 3 per heartbeat, the sounds undergo changes in intensity and quality. As the cuff pressure approaches the diastolic, the sounds often become dull and muffled quite suddenly and finally cease. The point of complete cessation is the best index of diastolic pressure. Record the subject's diastolic pressure.
5. Recording blood pressure. Blood pressure is recorded with the systolic over the diastolic:

   118/74 is 118 over 74. Record your blood pressure in the "Pulse and Blood Pressure" table.
6. Rotate members of your lab group so that each of you will have an opportunity to use the apparatus.
7. Optional: Variability in blood pressure.
   Exercise test (three or four students may work together on this experiment).
   a. Determine blood pressure from the one arm and pulse from the other arm with the subject sitting. Record on data sheet.
   b. Have the subject exercise vigorously for 1-2 minutes.
   c. Take the blood pressure and pulse immediately after exercise.
   d. Continue taking pulse every 2 minutes until the readings return to normal or stabilize.
   e. Record your results and plot a graph.

Parts III and IV: Data

Contraction of the ventricles exerts pressure on the blood vessels and is called systole. **Systolic pressure** is the pressure of the blood during ventricular contraction. This pressure **normally ranges between 110-140**. Following contraction the heart relaxes and blood rushes in from the pulmonary veins and venae cavae, filling the atria. This is diastole and this lower pressure is the **diastolic pressure**. This **ranges normally from 60-90**.

Blood Pressure is recorded: **systolic/diastolic**. For example, a person's blood pressure might be 115/76.

Guidelines on blood pressure interpretation are:

- ❖ 140/90 or above = hypertension (high blood pressure)
- ❖ 120/80 or above = prehypertension
- ❖ 120/80 or below = normal.

The guidelines are intended to urge people to exercise, reduce salt intake, limit alcohol drinks to two per day, and modify their behavior to avoid damage to arteries, thus increasing the risk of heart attack. An increase in blood pressure from 115/75 to 130/85 doubles the risk of death from heart disease.

What is the normal range for:

- ❖ systolic blood pressure? ______________
- ❖ diastolic blood pressure? ______________

Name ______________________________ Section ______________

Table 29.4
Pulse and Blood Pressure

Subject Name	Pulse	Blood Pressure	Interpretation[a]
1			
2			
3			
4			

[a] Interpret as: hypertension, prehypertension, normal.

Table 29.5
Blood Pressure and Pulse Following Exercise

		Normal	Time After Exercise					
			0 Min	2 Min	4 Min	6 Min	8 Min	10 Min
Blood Pressure	Systolic							
	Diastolic							
Pulse								

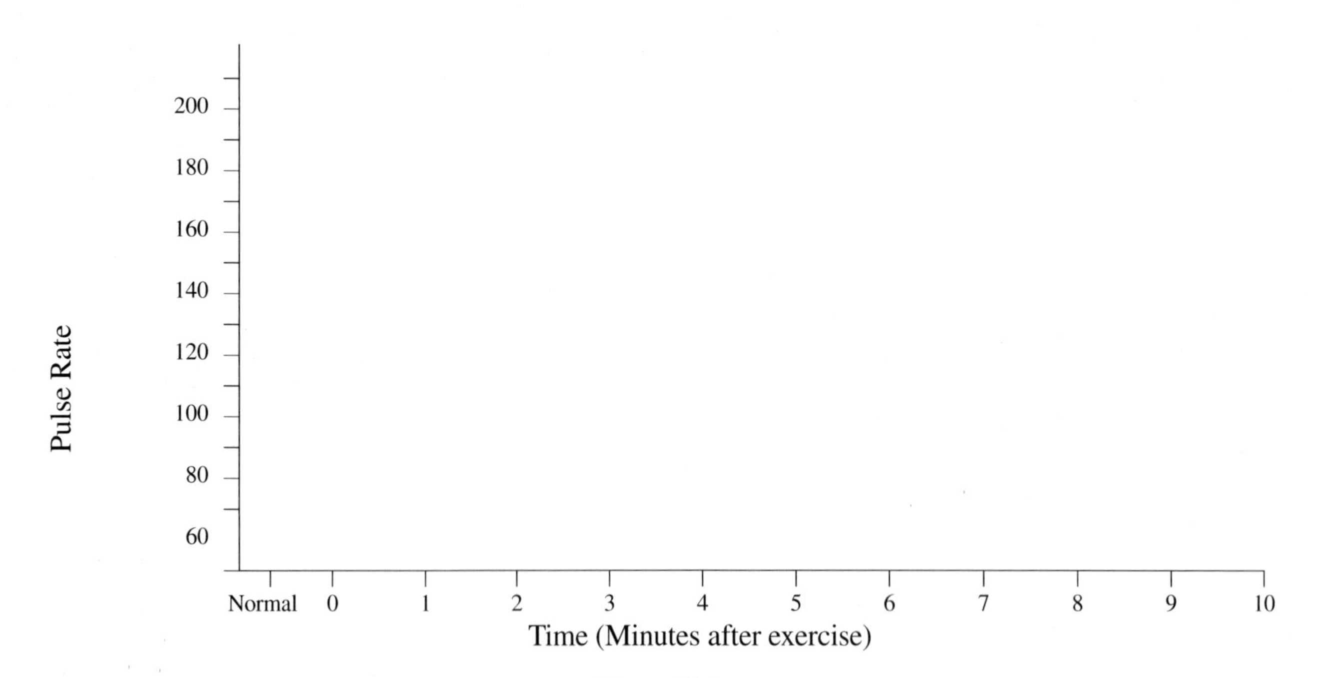

Figure 29.3
Blood Pressure and Pulse Following Exercise

Name ______________________ Section ______________

Lab 30

Animal Structure: Dissection of the Fetal Pig

Note: Terms in bold print should be identified on both the fetal pig and human manikin.

Problem

What is the appearance and physical relationship of the respiratory, digestive, excretory, and circulatory systems of the fetal pig and human?

Objectives

After completing this lab, the student will be able to:

1. Take a written or oral laboratory practical quiz identifying the major structures of the respiratory, digestive, excretory, and circulatory systems of the fetal pig and human. These structures are printed in bold.
2. List the functions of the specified structures.

Preliminary Information

This laboratory work will deal with the anatomy (structure) and physiology (function) of some parts of four major organ systems that are most directly concerned with the maintenance of life. The organs and their relationships in the fetal pig are the same as those found in most mammals, including humans.

The pig normally has a 17-week gestation period. The fetal pig you will be studying is a bypropduct of the meat processing industry. In any group of pigs sent to market, some will be pregnant. The fetal pigs cannot be sold as food. These fetal pigs were processed by a biological supply company, then purchased by the school. These fetal pigs were not killed so you would have a dissection specimen.

Whenever you are working with the anatomy of any organism, certain terms are commonly used. Many of these are listed here and you should know and be able to use all of them.

anterior—the head region
posterior—the tail region or end region
dorsal—the back side; up
ventral—the belly side; down
caudal—the tail region
lateral—the right or left side

medial—the middle or center
proximal—closer to the middle
distal—farther from the middle
pectoral—the shoulder region
pelvic—the hip region
right and left—the pig's right and left

Name ______________________________ Section ______________

Note: A **lab practical quiz** on fetal pig and human anatomy will be given on this lab. All parts identified by **BOLD PRINT** should be known on both **pig** and **human torso models**.

Part I: External Observation

Working in group of two, obtain a dissecting pan, two strings about the length of your arm, and a plastic bag from the supply area.

Obtain a fetal pig from the supply area.

After you have finished dissecting, place the pig and string in the plastic bag, tie, label with both partner's names written in pencil or permenent marker on masking tape, and place upright in the receptacle that is provided for your lab section.

1. During the external examination of the pig you are to observe the position of certain anatomical features. Observe these body regions: head, neck, trunk with two pairs of appendages, and tail. As in all mammals, the pig's trunk is divided into a **thorax**—the area from the anterior end of the pectoral region to the posterior end of the rib cage, and **abdomen**—the area from the posterior end of the rib cage to the posterior end of the pelvic region. The thoracic region contains the heart and lungs. The abdominal region contains the digestive, excretory, and reproductive organs.
2. On the ventral surface, there is a large **umbilical cord**. Three blood vessels pass through this cord. Two smaller thick-walled umbilical arteries and one large thin-walled umbilical vein carry oxygen, nutrients, and waste products between the fetal pig and its mother's placenta (the area of nutrient and waste exchange between fetal pig and mother). The umbilical arteries carry blood with waste products away from the fetal pig to the placenta. The umbilical vein carries oxygen and nutrients from the placenta to the fetal pig.

Part II: Organs of the Abdominal Cavity

Note: Dissection instructions should be closely followed and no organs should be removed unless directed.

Place the pig ventral side up in the dissecting pan as shown in Figure 30.1. To keep the pig in this position, tie the string to one front leg, run the string under the pan and tie it to the other front leg. Do the same for the back legs.

Use a scissors only to open the abdominal and thoracic cavities. While pulling the umbilical cord away from the body, make a left-to-right cut through the body wall one centimeter anterior to the umbilical cord. When the abdominal cavity has been reached, insert one blade of the scissors into the cavity and continue to open the abdominal and thoracic cavities following the pattern in Figure 30.1. Cut 1-2 will involve splitting the sternum.

Pick up the end of the umbilical cord and raise it slightly. Notice that it is attached by a vessel to one of the visceral organs. This is the **umbilical vein**. *Tie a string around this vein so it can be located later, and clip the vein between the string and umbilical cord.* Pull the umbilical cord posteriorly and let it rest between the hind legs. The abdominal cavity is now exposed, as in Figure 30.2. The structures within the coelom or body cavity should now be gently washed with a stream of tap water to remove any clotted blood, and preservative.

Cut off the flaps of the body wall on both sides of the pig. Cut through the ribs along the right and left sides of the pig to remove the rib cage. This will allow access to the thoracic cavity. Discard all flaps.

Name ______________________ Section ______________

With the aid of Figures 30.3 and 30.4 identify and study:

1. The **coelom** or body cavity which is divided into thoracic and abdominal cavities by the thin **diaphragm**.
2. The **peritoneum** lining the abdominal cavity and covering its **viscera** or internal organs. **Mesentery** is the double layer of peritoneum surrounding certain abdominal organs and attaching to the dorsal abdominal wall. Mesenteries are most easily located surrounding and holding the small intestine and its blood vessels.
3. The **liver**, a reddish-brown four-lobed structure posterior to the diaphragm.
4. The **gallbladder**, a membranous bag or sac the size of a pea partially imbedded in the dorsal side of the right medial lobe of the liver near its posterior border.
5. The **cystic duct**, which leads out of the gallbladder, joins the **hepatic ducts** from the liver to form a **common bile duct** that enters the duodenum. The duodenum is the first portion of the small intestine. *Use the torso model for these ducts.*
6. Thc **stomach** with the esophagus entering the cardiac portion and ending with the constructed pyloric portion.
7. The **spleen**, a flat elongated reddish-tan structure that lies just to the left of the stomach.
8. The light-colored, granular **pancreas** lying slightly dorsal and posterior to the stomach. Lift the stomach and small intestines to examine the pancreas. Use paper towels to blot up any excess fluids.
9. The much coiled **small intestine** attached to the stomach at the stomach's pyloric sphincter, a muscular valve. Raise the small intestine near the middle and examine the mesentery and blood vessels. The small intestine is usually subdivided into three sections. The duodenum is the first 20 cm in an adult, jejunum is the middle section and the last one-third is the ilium, which connects to the colon.
10. The **large intestine**, which follows the small intestine. There are several distinct parts of the large intestine:
    a. **The colon**, the anterior coiled part.
    b. **The rectum**, the somewhat straight posterior portion that opens to the outside through the **anus**.
    c. **The cecum**, the blind saclike beginning of the large intestine, at the juncture of the small and large intestines. In humans, the **appendix** is attached to the cecum.
11. The long **urinary bladder** extends ventro-posteriorly from the umbilical area. The right and left umbilical arteries lie along the sides of the bladder. In the initial dissection the umbilical cord and urinary bladder were pulled back between the pig's hind legs.
12. The **kidneys**, large bean-shaped organs lying against the dorsal wall in the middle of the abdominal cavity. The kidneys are easily seen by lifting the intestines.
13. The **ureter**, a tube carrying urine from the kidney to the urinary bladder.

Part III: Organs of the Thoracic Cavity

In the thoracic cavity observe:

1. The **pleural membrane**, the thin shiny membrane lining the cavity and covering the organs in it.
2. The **pericardium**, the sac that surrounds the heart.
3. A portion of the whitish **thymus gland** is attached to the pericardium. The rest of the thymus gland will be located later in the neck region.
4. The **lungs**, which lie on each side of the heart. *Remove and discard the left lung.*
5. The **esophagus**, which can be located anterior to the diaphragm. By tearing much of the pleural membrane between the lobes of the lungs and pulling the lobes to one side, three tubes can be seen: **posterior vena cava**, the most ventral tube; **esophagus**, the middle tube; **dorsal aorta**, the most dorsal tube.

Name ______________________________ Section ______________

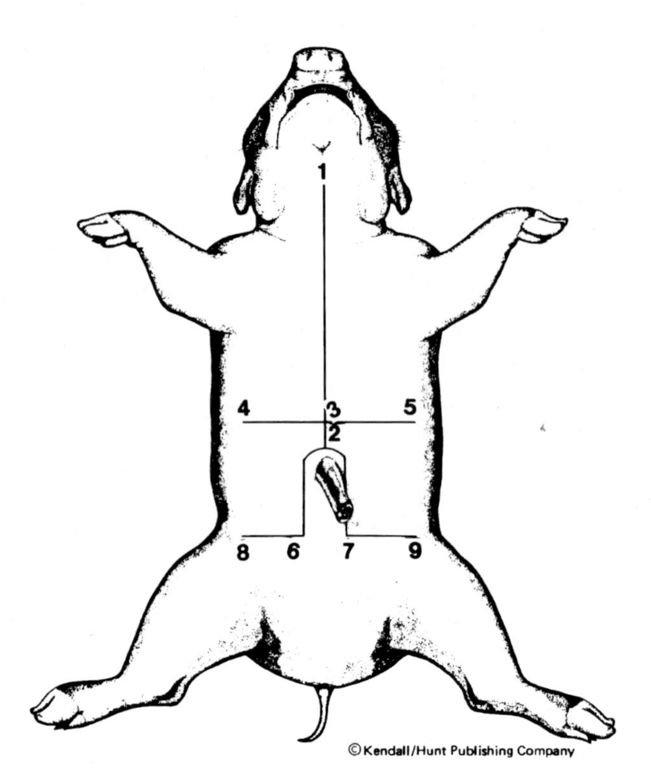

Figure 30.1
Dissection Cuts

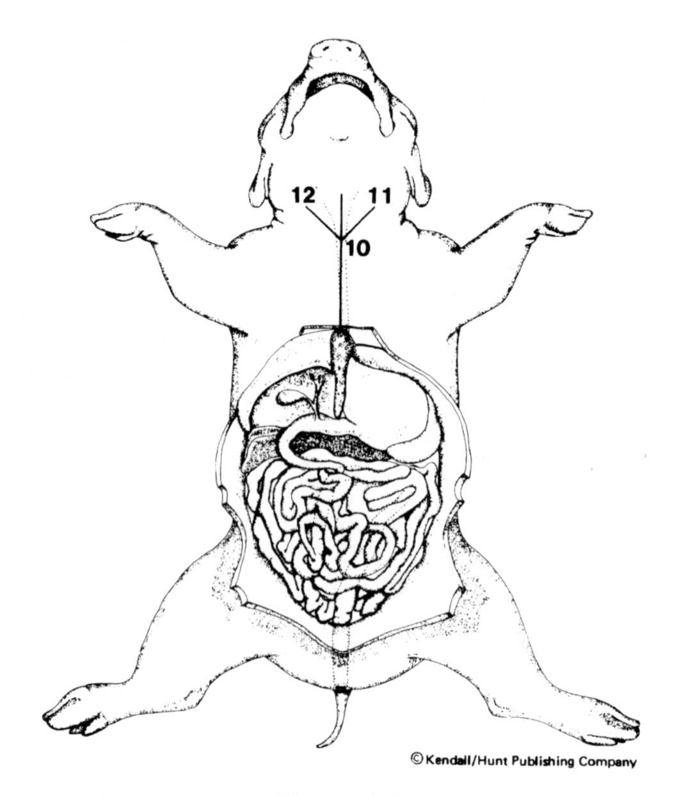

Figure 30.2
Abdominal Cavity

Name ____________________ Section ____________

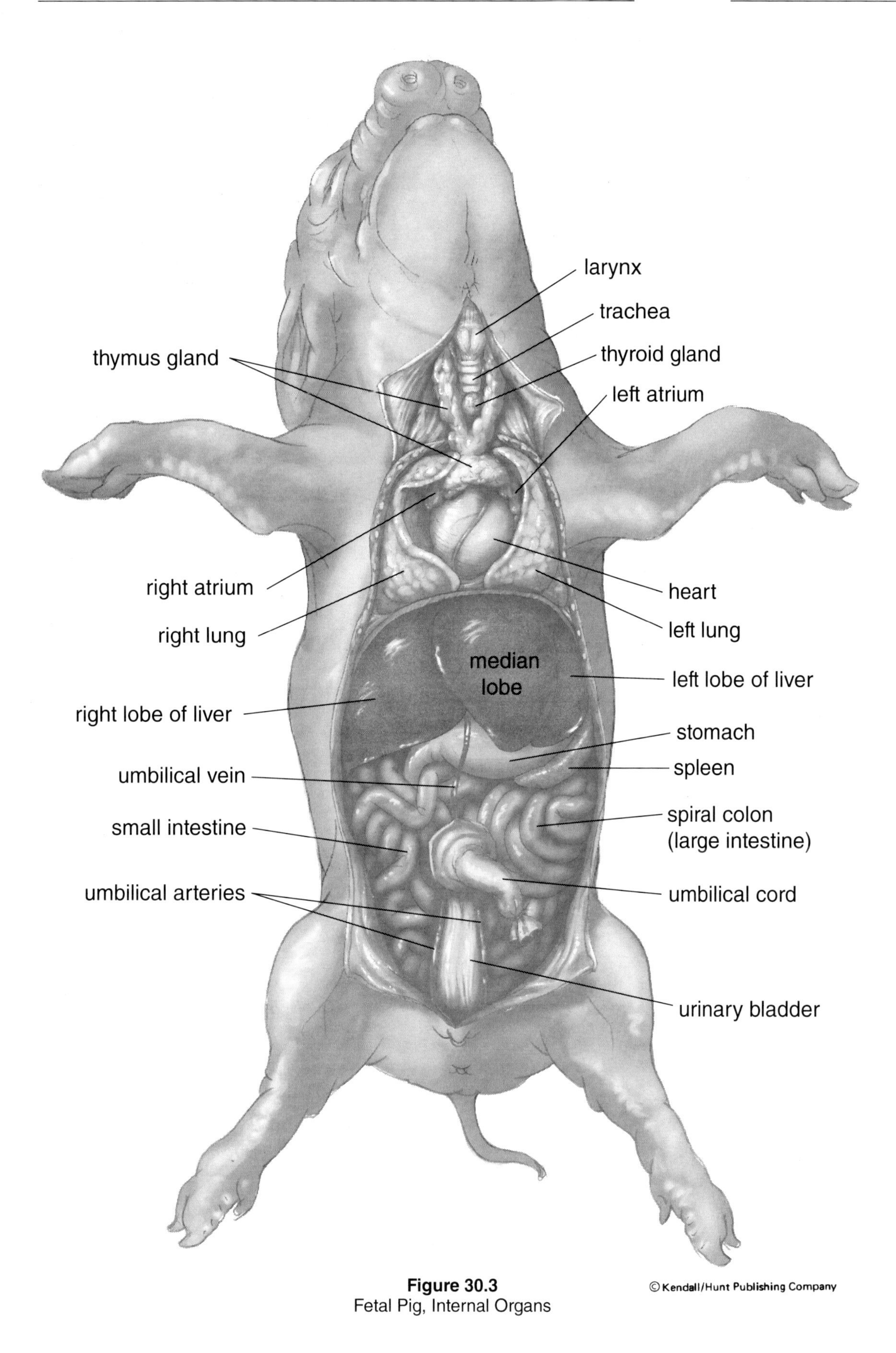

Figure 30.3
Fetal Pig, Internal Organs

Name ______________________ Section ______________

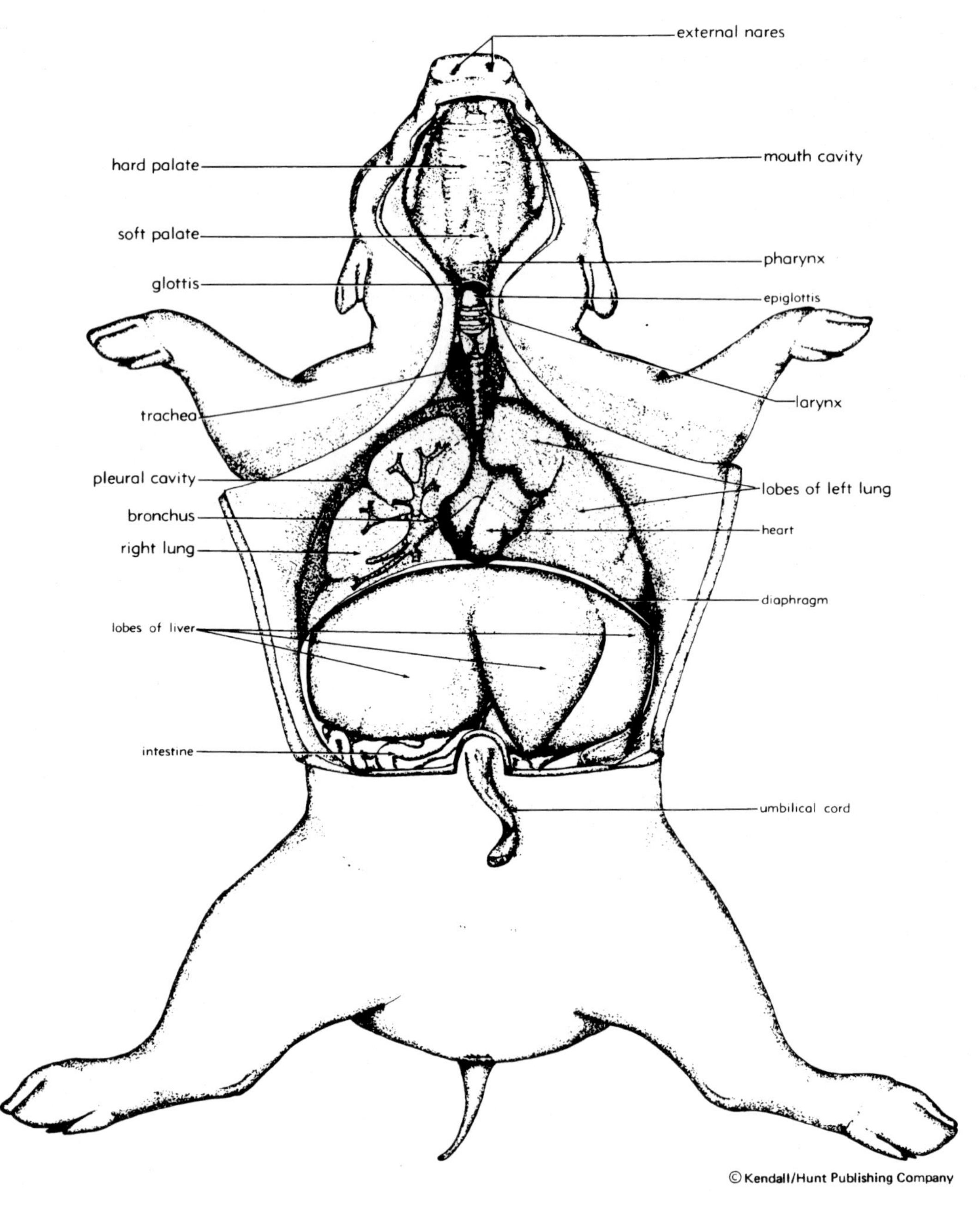

Figure 30.4
Fetal Pig, Respiratory System

Part IV: Structures of the Neck Region and Mouth

Using a blunt probe, carefully separate the structures in the neck region and observe:

1. The **larynx**, a large, firm white oval structure near the anterior portion of your incision.
2. The **trachea**, a tube leading from the larynx into the lungs. The trachea can be distinguished by cartilage rings, which keep it from collapsing.
3. The **thyroid gland**, a small reddish ball-like structure lying on the trachea slightly posterior to the larynx.

4. The **thymus gland**, a very large whitish globular structure running down each side of the neck, with a portion of it covering part of the pericardium. This gland is in close physical contact with the neck muscles. In adult humans, this gland is no longer present.
5. The **esophagus**, a tube just dorsal to the trachea. The esophagus carries food from the pharynx to the stomach.

Using a scissors, cut through the muscles on each side of the cheek by putting one point of the scissors in the pig's mouth so that the lower jaw can be pulled posteriorly. The bones of the lower jaw may need to be pulled or cut from their attachment to the skull. The lower jaw should flop down onto the pig's chest. Observe:

6. The **hard palate**, the washboardlike "roof" of the mouth.
7. The **soft palate**, posterior to the hard palate.
8. The flaplike **epiglottis**, which can fold back to cover the opening of the larynx during swallowing.
9. The **glottis**, which is the opening of the larynx.
10. From the mouth, insert a blunt probe down the **trachea** and later down the **esophagus**. Then feel for the tip of probe in the neck region.

Part V: Circulatory System

Remove the pericadium and attached thymus gland to expose the heart and associated blood vessels. Refer to Figures 30.5, 30.6, and 30.7 and locate the blood vessels listed below.

ARTERIES—carry blood away from the heart. Arteries should come injected with red latex.

1. **Dorsal aorta:** Large artery located just ventral to the vertebral column.
2. **Aortic arch:** The aorta leaves the left ventricle anteriorly then arches, becoming the dorsal aorta.
3. **Pulmonary artery:** Leaves the right ventricle and goes to the lungs. Most prominent vessel on top of the heart.
   Note: As you examine the fetal pig, the aortic arch is "behind" the pulmonary artery.
4. **Brachiocephalic or innominate artery:** First, very short branch off the aortic arch. Brachiocephalic artery immediately branches into the right subclavian artery and the bicarotid trunk, which branches into the right and left common carotid arteries.
5. **Right subclavian artery:** Runs into the right shoulder. *Careful*: The right subclavian artery is under some important veins.
6. **Left subclavian artery:** Second branch off the aortic arch; runs into the left shoulder.
7. **Right and left common carotid arteries:** Branches off the bicarotid trunk and goes into the head. Common carotids will branch to form internal and external carotid arteries. The jugular veins are beside the carotid arteries.
8. **Right and left renal arteries:** Branches off the dorsal aorta running to the kidneys.
9. **Right and left genital arteries:** Small vessels branching from the dorsal aorta just posterior to the renal arteries; run to testes or ovaries.
10. **Right and left external iliac arteries:** Large branches from the dorsal aorta running into the back legs.
11. **Right and left umbilical arteries:** Large arteries lying on each side of the urinary bladder. The first part of the internal iliac arteries forms the connection between the dorsal aorta and the umbilical arteries.

Name ______________________________ Section ______________

VEINS—carry blood toward the heart. Veins should come injected with blue latex.

12. **Anterior vena cava:** Also called **superior vena cava**, precava or precaval vein; short large blood vessel entering the right atrium bringing blood from the head; formed by joining of innominate veins.
13. **Posterior vena cava:** Also called **inferior vena cava**, post cava or postcaval vein; blood vessel entering the right atrium bring blood from the posterior regions.
14. **Right and left subclavian veins:** Branch from the shoulders entering the innominate veins.
15. **Right and left external jugular veins:** Larger, more lateral of the two pair of jugular veins.
16. **Right and left renal veins:** Run from kidneys to the posterior vena cava; near the renal arteries.
17. **Right and left genital veins:** Small vessels running from testes or ovaries to the posterior vena cava.
18. **Right and left common iliac veins:** Large veins carrying blood from the back legs to the posterior vena cava.
19. **Umbilical vein:** Small vein returning blood to the fetal circulatory from the placenta. It was cut and tied with string during initial dissection.

HEART—

20. **Right atrium:** Also called right auricle; saclike right anterior portion of the heart receiving the large anterior and posterior vena cavae. The right atrium appears to sit on top of the heart.
21. **Right ventricle:** Large muscular right posterior portion of the heart below the right atrium.
22. **Left atrium:** Also called left auricle; saclike left anterior portion of the heart.
23. **Left ventricle:** Larger muscular left posterior portion of the heart below the left atrium.
24. **Pulmonary artery:** Large artery leaving the right ventricle, carries non-oxygenated blood to the lungs.
25. **Aorta:** Large artery leaving the left ventricle.

The following structures need to be located on the heart model only.

26. **Tricuspid valve:** Three-flapped valve between right atrium and right ventricle.
27. **Bicuspid valve:** Also called **mitral valve**; two-flapped valve between left atrium and left ventricle.
28. **Chordae tendineae:** Cords attaching tricuspid and bicuspid valves to their ventricles so the valves are not pushed into the atria by the pressure of blood at contraction.
29. **Papillary muscles:** Attach chordae tendineae to the ventricles.
30. **Pulmonary veins:** Four small veins entering the left atrium from the lungs carrying oxygenated blood.
31. **Semilunar valves:** Half-moon shaped valves in the pulmonary artery and aorta preventing backflow of blood into the heart.
32. **Coronary arteries:** Arteries lying on the surface of the heart supplying blood to the heart muscle.

Name ______________________________ Section ______________

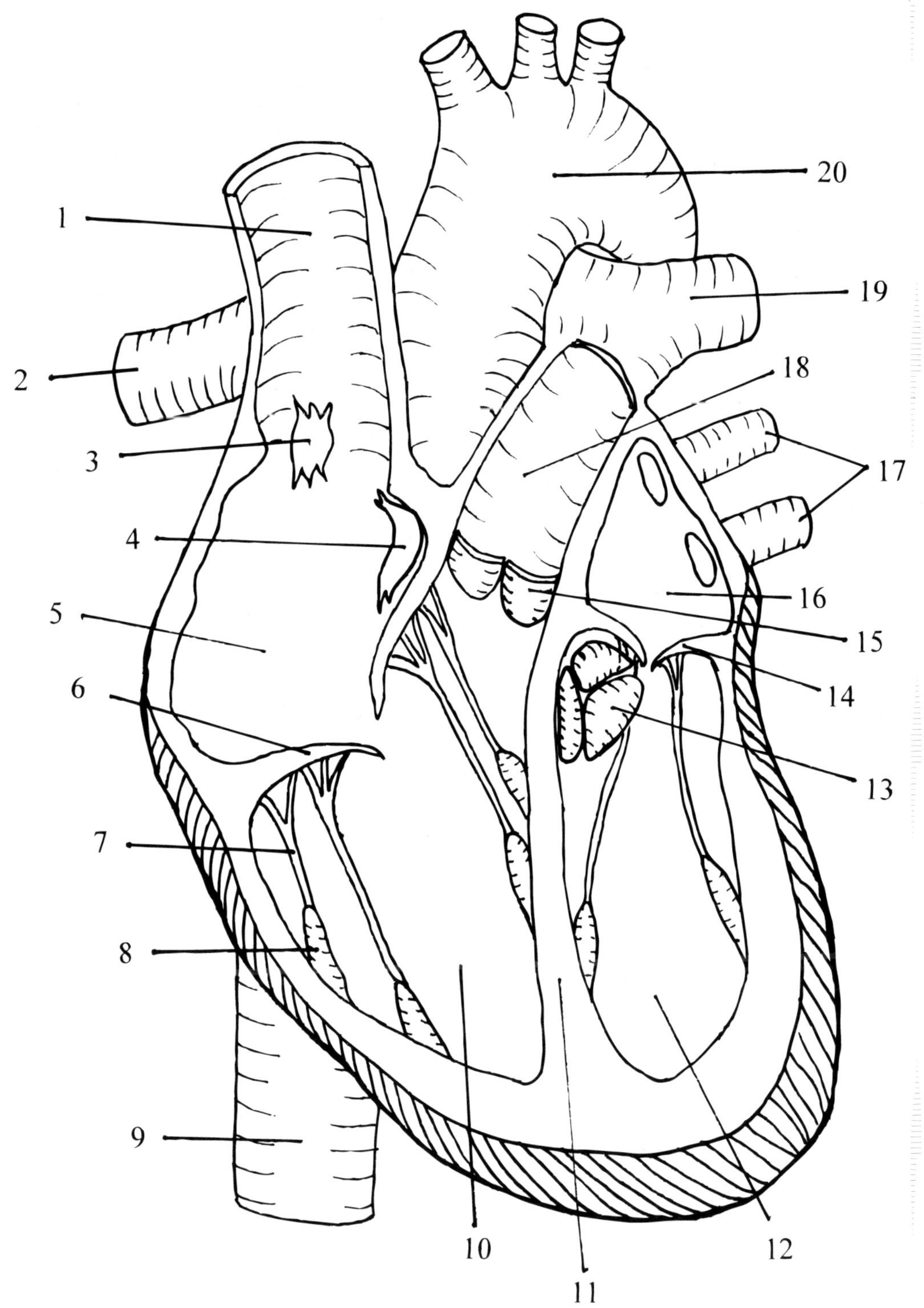

1. Anterior vena cava
2. Right pulmonary artery
3. Sinoatrial node
4. Atrioventricular node
5. Right atrium
6. Tricuspid valve
7. Chordae tendineae
8. Papillary muscle
9. Posterior vena cava
10. Right ventricle
11. Septum
12. Left ventricle
13. Semilunar valves of aorta
14. Bicuspid valve
15. Semilunar valves of pulmonary
16. Left atrium
17. Pulmonary veins
18. Pulmonary artery
19. Left pulmonary artery
20. Aorta

Figure 30.5
Human Heart

Name ______________________________ Section ______________

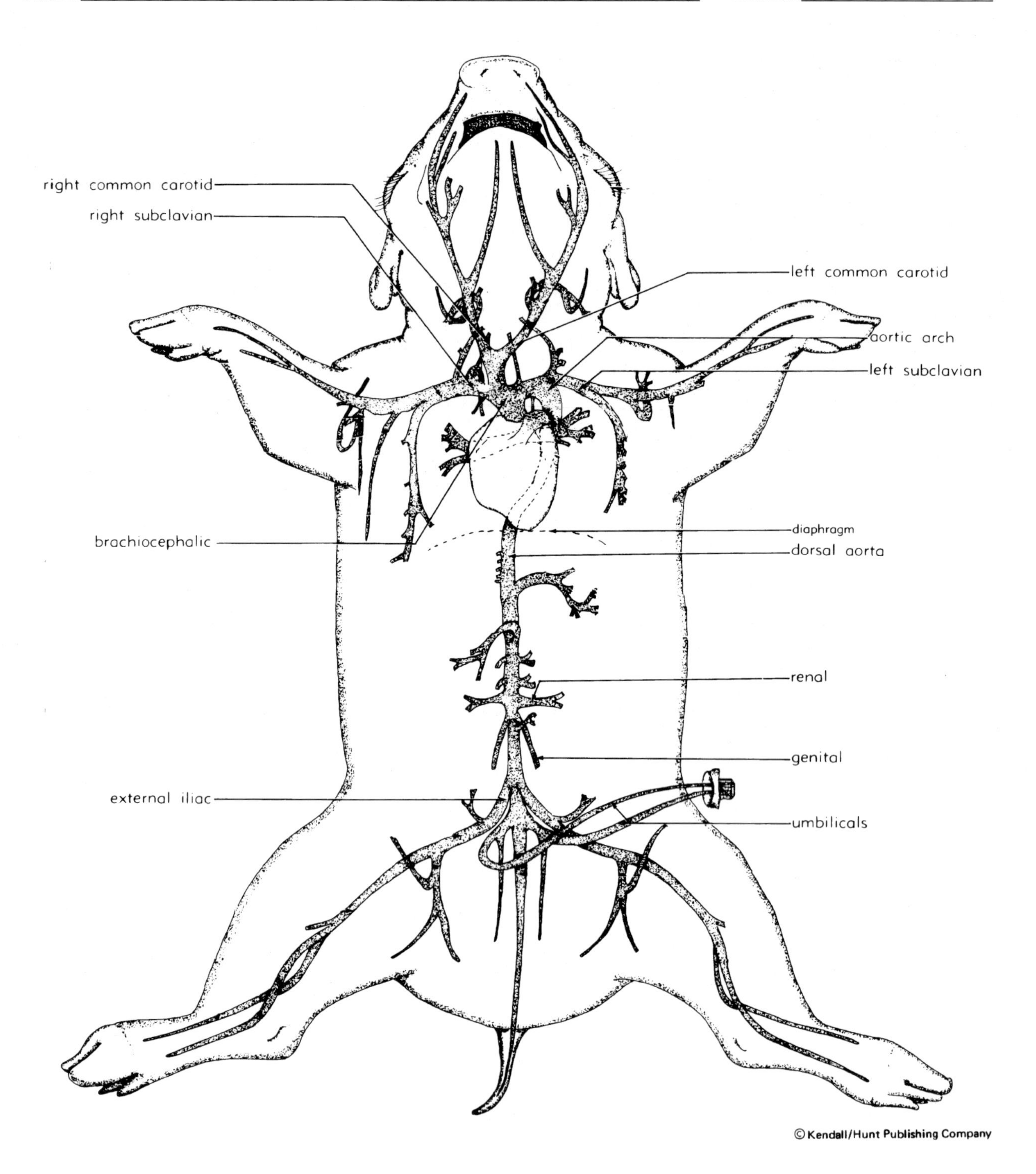

Figure 30.6
Fetal Pig, Arterial System

Name ______________________ Section ______________

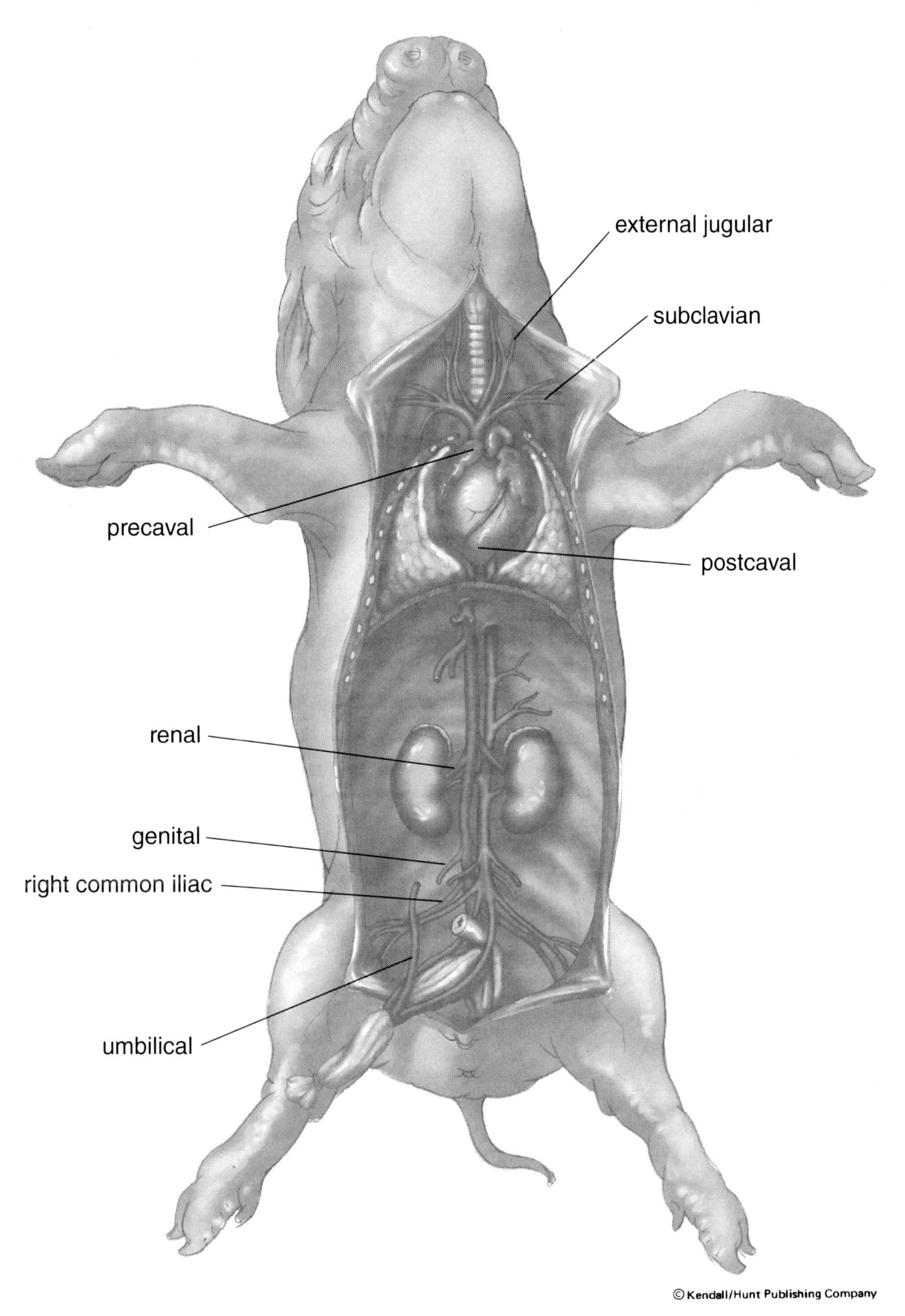

Figure 30.7
Fetal Pig, Venous System

Name ______________________ Section ____________

Part VI: Structures and Functions

Identify and give the functions of the following structures. Use the human heart for detailed and internal structures.

For the heart, indicate where blood comes from, where it goes and whether it is oxygenated or nonoxygenated.

Structures and functions from this section will be on the lab practical quiz using pigs, torsos, and models.

* Structures are only seen on pig.

1. Epiglottis: ______________________
2. *Thymus gland: ______________________
3. Thyroid gland: ______________________
4. Larynx: ______________________
5. Trachea: ______________________
6. Lungs: ______________________
7. Esophagus: ______________________
8. Stomach: ______________________
9. Small intestine: ______________________
10. Large intestine: ______________________
11. Appendix (in humans only): ______________________
12. Mesentery: ______________________
13. Diaphragm: ______________________
14. Liver: ______________________
15. Gallbladder: ______________________
16. Common bile duct: ______________________
17. Spleen: ______________________
18. Pancreas: ______________________
19. Kidneys: ______________________
20. Ureter: ______________________

Name ______________________________ Section ______________

21. Urinary bladder: ______________________________

22. *Umbilical arteries: ______________________________

23. *Pericardium: ______________________________

24. Heart: ______________________________

25. Right atrium: ______________________________

26. Right ventricle: ______________________________

27. Left atrium: ______________________________

28. Left ventricle: ______________________________

29. Tricuspid valve: ______________________________

30. Bicuspid (mitral) valve: ______________________________

31. Chordae tendinae: ______________________________

32. Papillary muscles: ______________________________

33. Semi-lunar valves: ______________________________

34. Coronary arteries: ______________________________

35. Pulmonary arteries: ______________________________

36. Pulmonary veins: ______________________________

37. Aortic arch/Dorsal aorta: ______________________________

38. Carotid arteries: ______________________________

39. Jugular veins: ______________________________

40. Subclavian arteries: ______________________________

41. Iliac arteries: ______________________________

42. Anterior (superior) vena cava: ______________________________

43. Posterior (inferior) vena cava: ______________________________

44. Renal arteries: ______________________________

Name ______________________________ Section ______________

Blood Flow Through the Heart

45. Trace the blood flow through the heart using arrows; use blue arrows to indicate nonoxygenated blood and red arrows to indicate oxygenated blood.
46. Label the indicated structures of the heart using Figure 30.8.

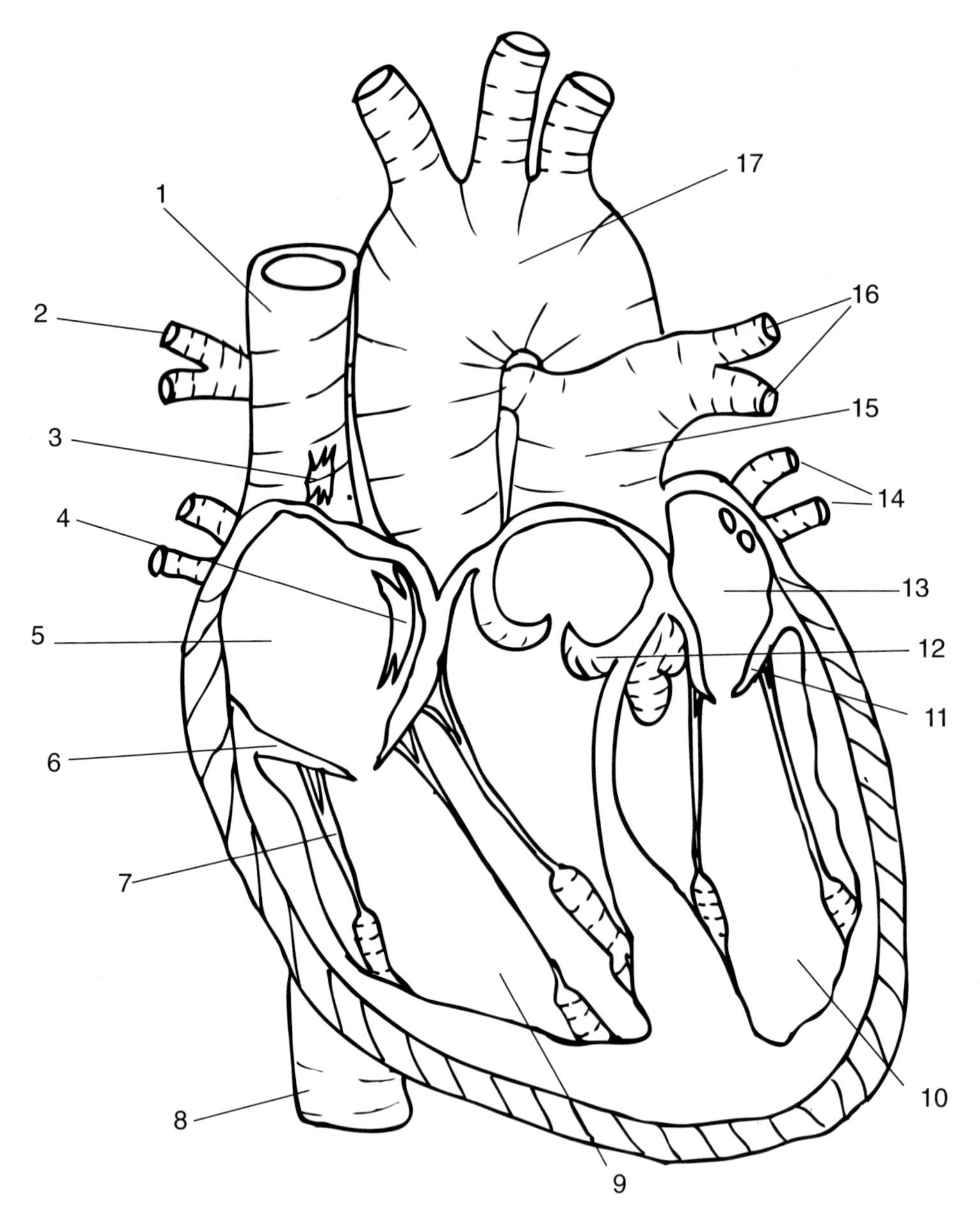

Figure 30.8
Human Heart

Name________________________________ Section______________

Lab 31

Kidney Function and Urinalysis

Developed by Sandra Gibbons

Problem

What can be learned about a person's health and kidney function by analyzing their urine?

Objectives

After completing this lab exercise, the student will be able to:

1. Explain the stages of urine production.
2. Use a Multistix SG reagent strip to determine the composition of a urine sample.
3. Discuss which urinalysis tests are abnormal and the probable cause behind the abnormality.
4. Identify the parts of a nephron.

Preliminary Information

The kidney is an important organ in the excretory system. It is responsible for eliminating toxic nitrogenous wastes produced during metabolism as well as controlling the water and salt concentration in the blood. The functional unit of the kidney is the nephron, which is actually a single long tubule associated with many blood vessels. Each kidney contains thousands of nephron tubules.

There are several stages to the production of urine by the nephron. The first stage is known as **filtration**. A cup-shaped swelling of the nephron known as the **glomerular capsule** surrounds a ball of capillaries known as the **glomerulus**. The glomerulus is much more permeable than other capillaries. The blood pressure in this region forces all small molecules from the blood and into the glomerular capsule. Filtration removes substances from the blood based on size. Small molecules such as water, glucose, amino acids, salts, urea, uric acid, and creatinine are forced out of the blood at this point. Large molecules such as proteins and blood cells remain in the blood.

Since many of the molecules removed during the filtration step are necessary for life (water, glucose, amino acids) they need to be taken out of the nephron filtrate and put back into the blood. This happens during the **selective reabsorption** stage, which occurs along the **proximal convoluted tubule** of the nephron. During this stage molecules such as water, glucose, amino acids, and salts move from the proximal tubule of the nephron back into the capillary network of the blood.

The next region of the nephron is known as the **loop of the nephron**. Along this portion water and salt will be moved out of the filtrate and back into the blood. **Tubular secretion** occurs at the portion of the nephron known as the **distal convoluted tubule**. Substances such as uric acid, creatinine, hydrogen ions, ammonia, and antibiotics such as penicillin are removed from the blood and actively transported into the nephron at this point. The fluid coming from the distal convoluted tubule will enter a **collecting duct**, which will collect the fluid (urine) produced by many nephrons and carry it to the renal pelvis portion of the kidney from where it will pass into the ureter going to the bladder and then to the urethra and finally out of the body.

The chemical analysis of urine (urinalysis) can tell the state of health of an individual. During this laboratory you will be testing five artificial urine samples using Multistix SG reagent strips, which will simultaneously test the glucose, bilirubin, ketone, specific gravity, blood, pH, protein, urobilinogen, nitrite, and leukocyte level in the urine. Table 31.2 lists some causes for the various levels you may observe.

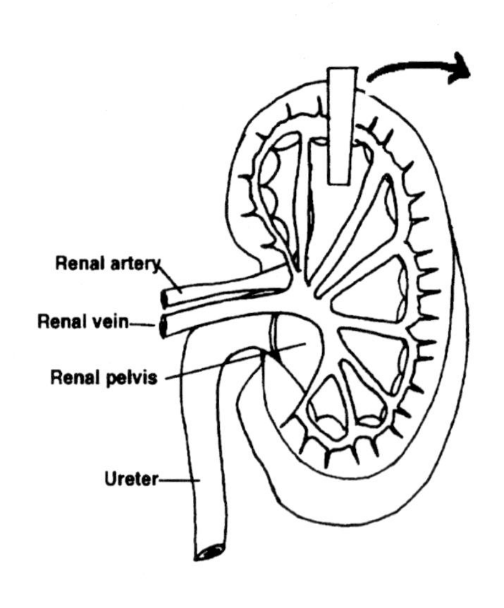

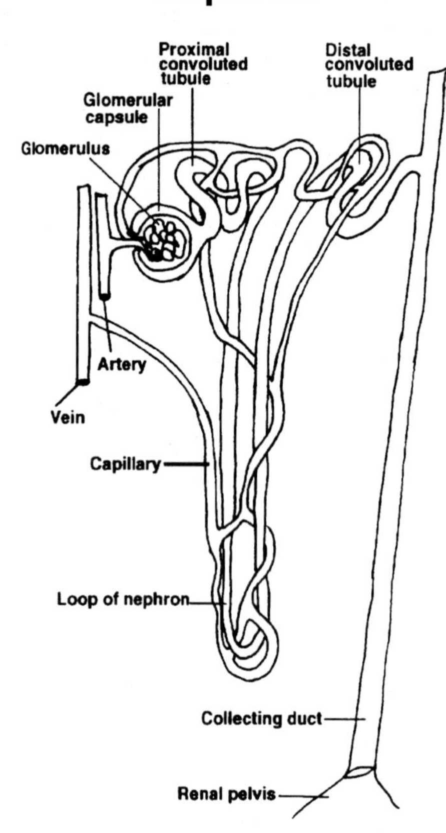

Figure 31.1
The Nephron

Name ______________________________ Section ______________

Part I: Methods and Materials

Work as a lab table.

1. Obtain a bottle of Multistix 10 SG reagent strips from the supply area. You need to familiarize yourself with the strips before you test any urine. The strip contains 10 colored squares, each of which tests for a different substance in the urine. The label on the Multistix container shows you which test each square represents as well as when you should read your results.
2. Obtain 5 test strips from the Multistix bottle and label them A, B, C, D, and E, using a permanent marker.
3. You will use a different reagent strip to test each of the urine samples. To conduct the test, use a forceps to dip the reagent strip entirely into the urine sample to insure that all squares are in contact with the urine. Remove the strip immediately to a paper towel and start timing. At the appropriate time, compare the strip with the colored squares on the bottle. When 30 seconds have elapsed, read the result for the glucose and bilirubin test; when a total of 40 seconds have elapsed read the ketone test; at 45 seconds read the specific gravity test; at 60 seconds read the blood, pH, protein, urobilinogen, and nitrite tests. At 2 minutes read the leukocyte test. Record the results in the table.
4. Optional: You may conduct these tests on your own urine.
5. Discard the reagent strips in the garbage.

Name ______________________________ Section ______________

Part II: Data

Table 31.1
Results of Urinalysis on Each Urine Sample A-E

Test	Samples					
	A	B	C	D	E	Student
Leukocyte normal: negative						
Nitrite normal: negative						
Urobilinogen normal: 0.2-1						
Protein normal: negative to trace						
pH low: <4.5 high: >8.0						
Blood normal: negative to trace						
Specific Gravity low: <1.010 high: >1.025						
Ketone normal: negative						
Bilirubin normal: negative						
Glucose normal: negative						

Name ______________________ Section ______________

Table 31.2
Possible Causes for Abnormal Test Results

Test	Results	Medical Cause	Other Cause
Leukocyte	Moderate–Large	Infection	None
Nitrite	Positive	Infection	None
Urobilinogen normal 0.2-1	1–8	Liver disease (hepatitis, cirrhosis)	Red blood cell destruction
Protein	30–2000	Severe anemia or infection	High protein diet or heavy exercise
pH	Low <4.5 High >8.0	Uncontrolled diabetes Severe anemia	High protein diet or cranberry juice in diet Diet rich in vegetables or dairy products
Blood	Small–Large	Infection	Menstruation
Specific gravity	Low <1.010 High >1.025	Severe kidney damage Uncontrolled diabetes or severe anemia	Increased fluid intake Decreased fluid intake or loss of fluid
Ketone	Small–Large	Uncontrolled diabetes	Produced during fat metabolism, could be due to dieting or high protein diet
Bilirubin	Small–Large	Liver disease (hepatitis)	Bile duct blocked
Glucose	250–2000	Uncontrolled diabetes	A large meal eaten recently

Part III: Discussion and Conclusions

1. Discuss the results of each sample including the probable medical problem associated with each sample. Give a diagnosis; don't re-write the data.

   A

   B

   C

   D

   E

Name ______________________________ Section ______________

2. At which point along the nephron would glucose be entering into the filtrate (urine)?

3. At which point along the nephron would glucose be reabsorbed into the blood?

4. At which point along the nephron would an antibiotic enter the filtrate (urine)?

5. You are being treated with antibiotics for a urinary tract infection. How could you use the Multistix to see if the antibiotics are working? Be specific about which tests you would use.

Name ______________________ Section ____________

Lab 32

Urogenital Systems

Problem

What is the appearance and physical relationship of the excretory and reproductive systems of the human?

Objectives

After completing this lab exercise, the student will be able to:

1. Take a laboratory practical quiz identifying the major structures and functions of the excretory and reproductive systems of the human. These structures are printed in bold.
2. List the functions of the specified structures.
3. Identify the parts of a sperm cell.
4. Identify the parts of a follicle in the ovary.

Preliminary Information

This laboratory work will deal with the anatomy (structure) and physiology (function) of the excretory and reproductive systems of the human.

For human anatomy, several types of models are available. Study all of the available models since some models show some parts better than others.

Refer to Figures 32.1 through 32.4.

Part I: Excretory System

Note: Terms in bold print should be identified on all models and labeled on Figures 32.1 through 32.4.

1. The **urinary bladder** is located ventro-posteriorly in the human pelvis. The urinary bladder collects and temporarily stores urine.
2. The **kidneys** are large bean-shaped organs lying against the dorsal wall in the middle of the abdominal cavity. The kidneys can be seen only on the torso model. The kidneys help maintain homeostasis by cleaning the blood of liquid wastes (such as urea) and maintaining proper salt balance in the blood.
3. The **ureter** is a tube draining the urine from the kidney to the urinary bladder.

Name ______________________ Section ______________

Part II: Reproductive Systems

Female Reproductive System

1. The ovary appears as a small oval structure in the abdominal cavity below the kidney. There is one ovary on each side. The ovary produces eggs and hormones.
2. An **oviduct** or **Fallopian tube** or **uterine tube** is a tube beginning near the surface of the ovary then going to the middle of the body where it connects to the uterus. The oviduct transports eggs to the uterus. Fertilization usually occurs in the oviduct.
3. The **uterus** is located at the medial ends of the oviducts. The uterus runs dorsal-anterior to the **urinary bladder**. The uterus holds and nourishes the developing baby.
4. The **endometrium** is the lining of the uterus that builds then deteriorates on a monthly cycle. During pregnancy, it thickens to provide increased blood flow and thus nourishment to the developing baby.
5. The **myometrium** is the smooth muscle portion of the uterus outside the endometrium. Its contractions expell the baby during labor and delivery.
6. The **cervix**, which is the lower part of the uterus, can be distinguished as a firm structure between the vagina and the body of the uterus. The cervix holds the developing baby within the uterus.
7. The **vagina** is located dorsal to the urethra. The vagina is the birth canal. At its inner end, the vagina connects to the uterus.
8. The **labia majora** are fleshy and more prominent folds of skin located externally to the small folds of skin, the **labia minora**. The labia protect and inhibit drying of the vagina.
9. The **clitoris**, a homolog to the penis, is located at the anterior junction of the labia majora. Its function is sexual stimulation.
10. The **urethra** carries urine from the **urinary bladder** to outside of the body.
11. The **pubic bone** or pubis is the ventral part of the pelvic (hip) bone. The **pubic symphasis** is the midline where the left and right pubic bones meet.

Name ______________________________ Section ______________

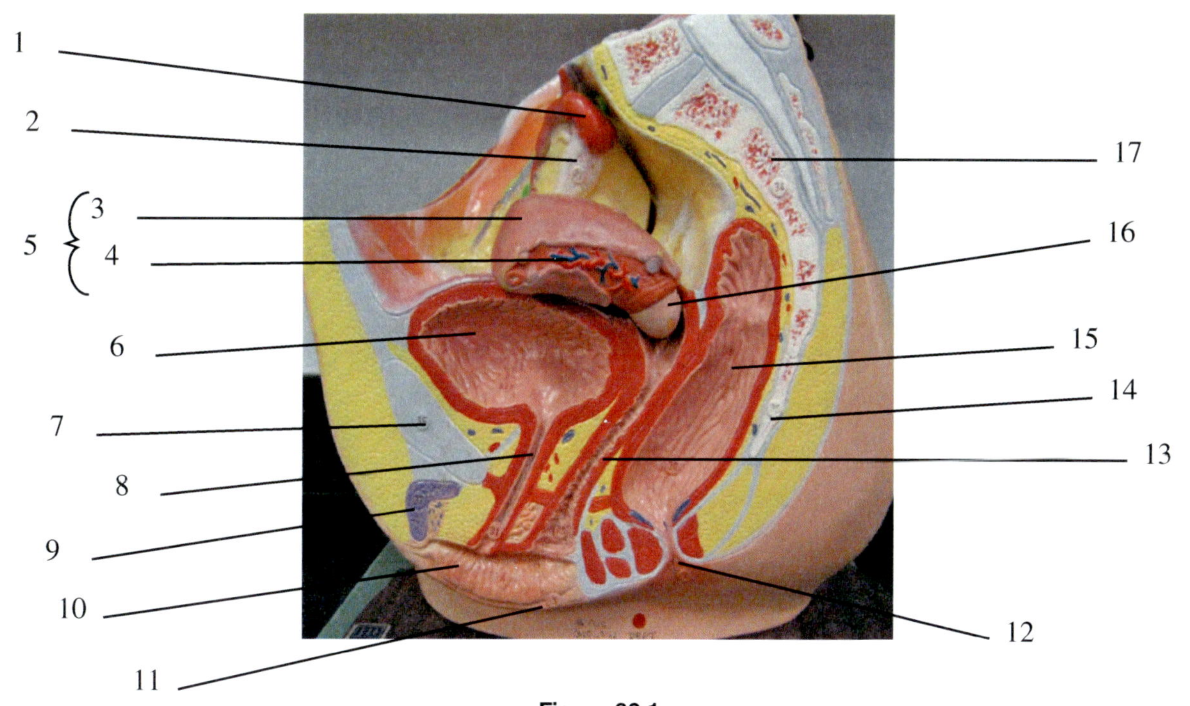

Figure 32.1
Female Urogenital System

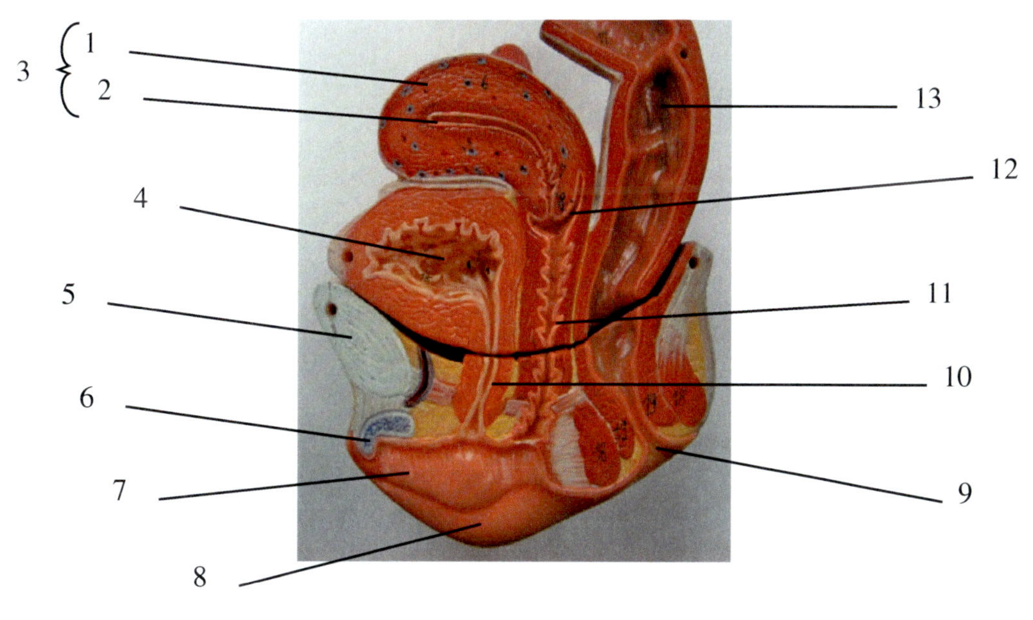

Figure 32.2
Female Urogenital System

Name ______________________ Section ______________

Male Reproductive System

1. The **scrotum** is the sac-like structure enclosing and suspending the testes and ducts outside the abdominal cavity.
2. The **testes** are paired, oval structures located in the scrotal sac. Sperm production and testosterone production occurs in the testes.
3. The **seminiferous tubules** are highly coiled tubules located in each of the testis. Sperm develop in these tubules.
4. The **epididymis** is a tightly coiled tubule lying on the surface of the testis. The epididymis connects to the vas deferens. Sperm maturation occurs in the epididymus.
5. The **vas deferens** or **ductus deferens** is the tube running through the inguinal canal out of each scrotal sac. The vas deferens from each testis can be seen inside the abdominal cavity where they empty into the urethra by joining together to form the **ejaculatory duct**. Vas deferens and the ejaculatory duct transport sperm.
6. The **seminal vesicles** are paired glands located dorsally where the **ejaculatory duct** forms. Seminal vesicles produce a sugar solutuion, which sperm use for energy.
7. The **prostate gland** is a gland located at the junction of the **ejaculatory duct** and the **urethra**. The prostate gland produces an alkaline (basic) solution to neutralize the acidity in the vagina so the sperm can survive.
8. The **Cowper's glands** or **bulbourethral glands** are small structures located on each side of the urethra at the bend in the urethra where it joins the penis. Cowper's glands produce a lubricating solution to aid in copulation.
9. The **urethra** runs through the penis from the urinary bladder to the outside.
10. The **corpus cavernosum** is erectile tissue running nearly the length of the penis. The corpus cavernosum is anterior to the corpus spongiosum.
11. The **corpus spongiosum** is the erectile structure located around the urethra in the penis.
12. The **foreskin** is the retractable skin covering the outer end of the penis. The foreskin is often removed by circumcision.
13. The **pubic bone** or pubis is the ventral portion of the pelvic (hip) bone. The **pubic symphasis** is the midline where the left and right pubic bones meet.

Name ______________________ Section ______________

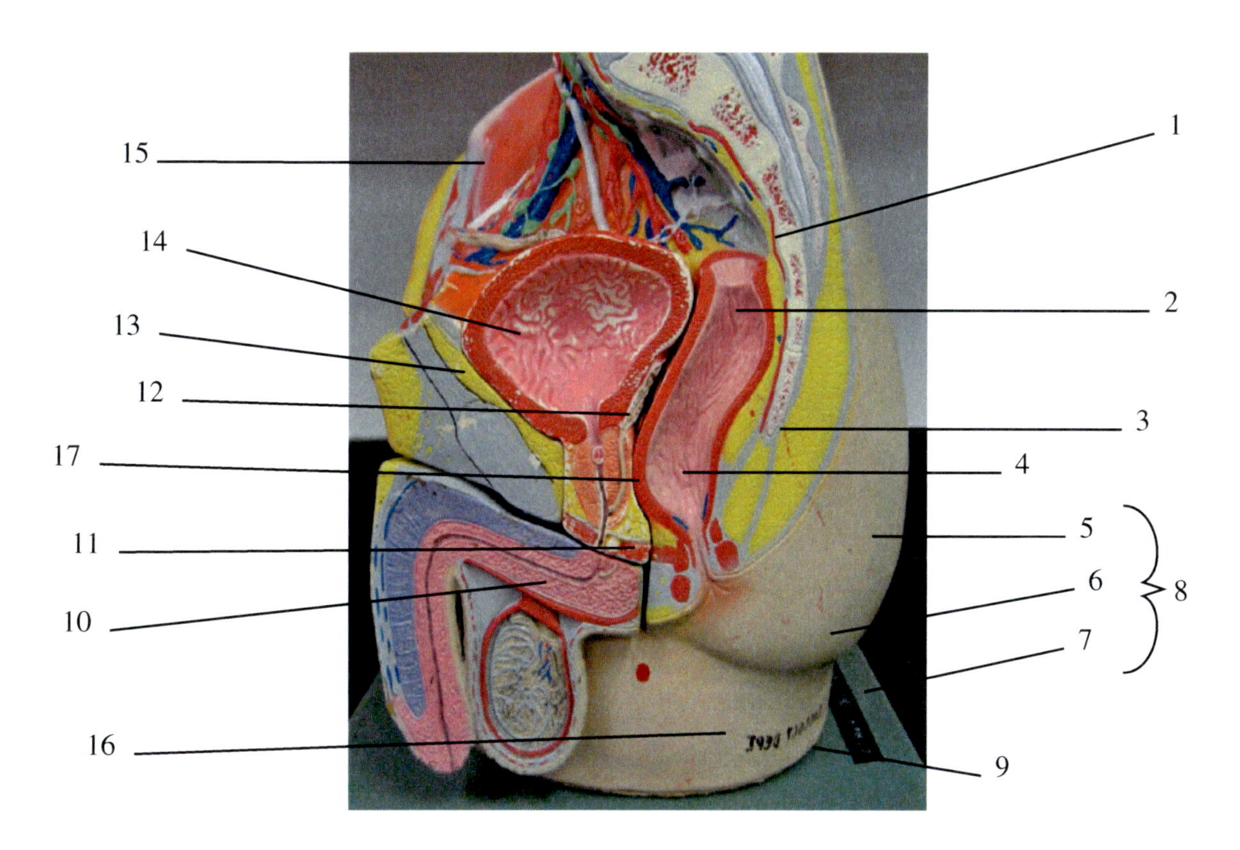

Figure 32.3
Male Urogenital System

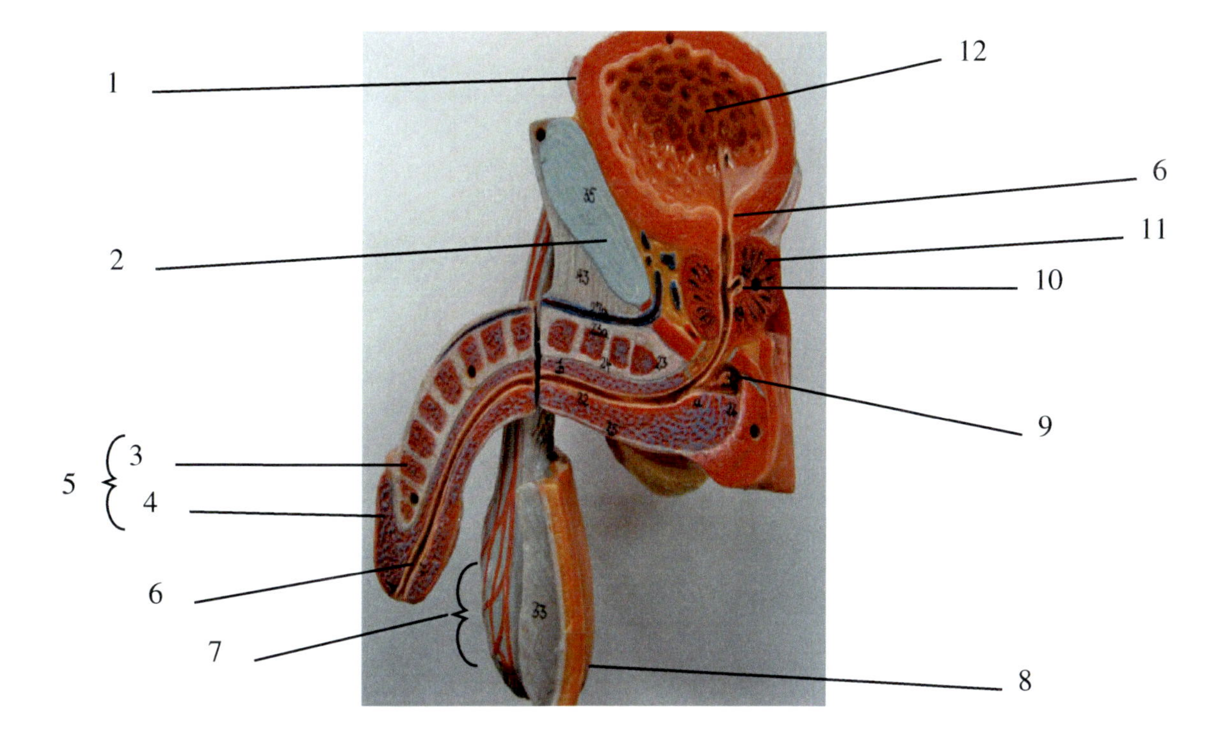

Figure 32.4
Male Urogenital System

Name______________________________ Section______________

Part III: Reproductive Physiology

Male

The **testes** are contained in the **scrotum**. Each testis contains meters of small hollow coiled **seminiferous tubules** in which the sperm cells develop by meiosis in a process called spermatogenesis. The sperm cells are moved out of the testis to the **epididymus** where they mature, are temporarily stored, and reabsorbed if in excess. **Interstitial cells** located in the spaces between the seminiferous tubules produce testosterone, a major male sex hormone.

1. Obtain a sperm slide from the supply area.
2. Examine the sperm cells under high (430X) power of a monocular microscope.
3. Diagram and label the:
   a. **nucleus**
   Note: The nucleus is capped by an acrosome, which contains enzymes involved in fertilization. It is not visible with a light microscope.
   b. **flagellum.**

Female

In the ovary, precursor egg cells (oogonia) are formed during fetal development. The oogonia mature into primary oocytes. These begin meiosis (oogenesis). In oogenesis, meiosis is not completed until after fertilization. The number of oocytes decreases with age. At puberty in humans, each ovary contains approximately 200,000 oocytes. This number is more than sufficient since only about 400 to 500 will complete development into mature eggs.

During each menstrual cycle, hormones from the pituitary gland stimulate the development of a **follicle (secondary oocyte and its surrounding support cells)**. As development proceeds, the follicular liquid increases in volume. The follicle moves to the surface of the ovary where the oocyte is released. Most of the follicle cells remain on the surface of the ovary forming a corpus luteum, a scab-like structure. After fertilization, which normally occurs in the upper portion of the oviduct, the secondary oocyte completes meiosis. Then the male chromosomes and female chromosomes restore the chromosome number to its full complement (46 in humans) and a zygote (fertilized egg) is formed.

1. Obtain an ovary slide from the supply area.
2. Examine the slide under scanner (4X) power.
3. Locate the large spaces in the ovary. These are follicles filled with fluid.
4. Increase the magnification to low (10X) power and observe a large follicle. A large cell, the secondary oocyte, is supported in the follicular fluid by a base.
5. Diagram and label the **secondary oocyte** supported by:
   a. the **base** holding the secondary oocyte in the **follicular liquid**,
   b. the **follicular liquid** around the oocyte,
   c. the **follicle cells** around the liquid.

Sperm	Ovary

Name ______________________________ Section ______________

Part IV: Structures and Functions

Identify and give the functions of the following structures.

Structures and functions from this section will be on the lab practical quiz.

1. Kidney: ______________________________
2. Ureter: ______________________________
3. Urinary bladder: ______________________________
4. Urethra: ______________________________
5. Ovary: ______________________________
6. Oviduct (uterine tube or fallopian tube): ______________________________
7. Uterus: ______________________________
8. Cervix: ______________________________
9. Endometrium: ______________________________
10. Myometrium: ______________________________
11. Vagina: ______________________________
12. Clitoris: ______________________________
13. Labia (majora and minora): ______________________________
14. Pubic bone: ______________________________
15. Scrotum: ______________________________
16. Seminiferous tubules: ______________________________
17. Epididymus: ______________________________
18. Vas deferens: ______________________________

Name ______________________________ Section ______________

19. Seminal vesicle: ______________________________

20. Prostate gland: ______________________________

21. Cowper's (bulbourethral) gland: ______________________________

22. Penis: ______________________________

23. Corpus cavernum/corpus spongiosum: ______________________________

24. Sperm cell: ______________________________

25. Secondary oocyte: ______________________________

26. Anus: ______________________________

27. Rectum: ______________________________

28. Sacrum: ______________________________

29. Coccyx: ______________________________

30. Spinal cord: ______________________________

Name ______________________ Section ____________

Lab 33

Early Embryological Development

Problem

What are the early embryological stages of development?

Objectives

After completing this laboratory exercise, the student will be able to:

1. Identify the following stages of development:
   a. fertilization
   b. cleavage
   c. blastula
   d. gastrula.
2. Note the trend in cell number and cell size as development proceeds from fertilized egg to blastula.
3. Name the three germ layers of the gastrula.
4. Identify the major portions of the body which develop from each germ layer.

Preliminary Information

The general pattern of reproduction in most higher types of animals is the same. Sex cells (eggs and sperm) are produced by meiosis. Sperm and egg join in a process of **fertilization** forming a zygote. In some organisms, fertilization is external, usually occurring in water. In other organisms, fertilization occurs in the reproductive tract of the female. Following fertilization, the **zygote** (fertilized egg) begins a series of mitotic cell divisions. These early embryological divisions are called **cleavage** since the large egg is successively divided or cleaved into smaller cells. For a while, the cells divide so quickly that they do not have time to grow between divisions. Consequently, each new generation of cells becomes progressively smaller. Finally the speed of mitosis decreases allowing the subsequent generations of cells time to grow between divisions.

Part I: Procedure

1. In this lab exercise, the early stages of development will be examined. Obtain a prepared microscope slide of Starfish development, w.m. from the supply area. The "w.m." means whole mount. In a whole mount preparation, the entire specimen is placed on the slide rather than a slice or section of a specimen.
2. Scan the slide to locate the following stages of development.
3. Draw each stage of development in the space provided.

Name ______________________________ Section ______________

Part II: Early Stages of Development

1. **Unfertilized egg**
   The unfertilized egg is a large single cell with a jelly-like coat around it. There is a distinct nucleus and a single nucleolus inside the nucleus.
2. **Zygote**
   In the fertilized egg (zygote), the nucleus and nucleolus disappear. Chemicals released from the zygote harden the jelly coat to prevent other sperm from entering.
3. **Two-cell stage**
   The single-celled fertilized egg pinches into two cells. Each new cell is half the size of the egg cell.
4. **Four-cell stage**
   Each cell of the two-cell stage pinches into two, forming a four-cell structure. Each new cell is half the size of its parent cell.
5. **Eight-cell stage**
   Each cell of the four-cell stage simultaneously divides.
6. **Sixteen-cell stage**
   Each cell of the eight-cell stage simultaneously divides.
7. **Morula or thirty-two-cell stage**
   The morula is a solid ball of 32 cells looking like a raspberry. Cell division proceeds so quickly that the new cells do not have time to grow.
8. **Blastula**
   The blastula is a hollow ball of cells with a central cavity called the blastocoel.
9. **Gastrula**
   The gastrula stage begins when the cells at one spot on the hollow ball stage begin to migrate into the central cavity (blastocoel). This appears as a depression on the surface of the blastula. Imagine pushing your finger into a tennis ball. This is the way the gastrula will appear. The infolding of the outer surface will continue until the blastocoel is eliminated. At this point, two of the three embryonic germ layers can be seen, the ectoderm and the endoderm. The different cell layers of the gastrula are called germ layers because they contain the initial cells, which will later develop into specific body structures. Soon the third germ layer, the mesoderm, will form between the first two layers.

   The **three embryonic germ layers** are:

   a. the **ectoderm** or outer layer
   b. the **endoderm** or inner layer, which has just infolded
   c. the **mesoderm** or middle layer, which forms between the ectoderm and endoderm.
      The ectoderm will form the skin and nervous system. The endoderm will form primarily the digestive tract. The mesoderm will form primarily muscle and bone.

Name ______________________________ Section ______________

Table 33.1
Early Embryological Development

Unfertilized egg	Zygote
Two-cell stage	Four-cell stage
Eight-cell stage	Sixteen-cell stage
Morula stage	Blastula stage
Gastrula stage	

Name ______________________________ Section ______________

Part III: Discussion

1. How can you distinguish an unfertilized egg from a fertilized egg?

2. As development proceeds from zygote to morula, what happens to the:
   a. number of cells?

   b. size of cells?

   c. size of the cell mass?

3. Construct a table listing the three germ layers of the gastrula and the major body portions formed from each layer.

Name ______________________ Section ______________

Lab 34

Human Reflexes and Senses

Problem

Using a variety of experiments, can you determine if the senses are always reliable?

Objectives

After completing this lab, the student will be able to:

1. Identify different reflex responses.
2. Discuss sensitivity to touch in different parts of the body.
3. Explain the effects of continuous exposure of a substance on the sense of smell and its subsequent reliability.

Preliminary Information

As we touch, smell, hear, and see things, we experience our environment. A loss of any one of the senses can be a major problem. People can adjust, of course, but the world is never the same.

Reflexes, like senses, are an important survival mechanism. They enable your body to respond to environmental changes quickly without your brain having to think about it. After the event has happened, such as pulling your hand off a hot stove, the brain becomes aware of the event. The ability to respond quickly can protect your body.

In this lab, you will explore some of the human reflexes and senses. Consider what would happen if these were no longer present.

Part I: Response Time and Light Intensity

In this experiment you will measure response time in bright light and dim light by measuring how quickly a meter stick can be grabbed.

1. Obtain a meter stick.
2. Your lab partner will catch the meter stick as it falls between his/her hands. Your partner should be standing with the palms of the hands 10 centimeters apart.
3. Hold the meter stick vertically with the zero mark even with the top of the hands.
4. Let go of the meter stick so it drops straight down. Your partner should catch it as quickly as possible.
5. Record the number where the meter stick was caught.
6. Repeat the experiment four more times, recording the data.

Name ______________________ Section ____________

7. Trade places and repeat. Record the data.
8. Turn the lights off or use a dimly lit room and repeat. Record all data.

Table 34.1
Reaction Speed vs. Light

	Experiment #1		Experiment #2	
	Light	**Dark**	**Light**	**Dark**
Trial 1				
Trial 2				
Trial 3				
Trial 4				
Trial 5				
Average				

❖ How does light intensity affect reaction rate?

❖ If you are driving a car at night, what adjustments should you make?

Part II: Touch Sensitivity

Working in pairs, determine areas of greater sensitivity by having the subject note when two pins can be felt as only one.

1. Obtain two dissecting pins and clean them with alcohol.
2. The subject should close his eyes while the experimenter tests the areas listed in the following table.
3. Each area is tested by the experimenter holding the pins 10 mm apart and very gently touching the subject with both pins at the same time. Move the pins closer together one mm at a time and touch the subject again. The subject should be told to indicate when both pins feel like one. Measure and record the distance between the pins in the table.
4. To interpret the data: **the smaller the number, the more sensitive the area.**

Name ______________________ Section ______________

Table 34.2
Sensitivity to Touch

Area Tested	Distance Between Pins
Index Finger	
Back of the Hand	
Palm of the Hand	
Back of the Neck	
Inner Forearm Near Elbow	

❖ It is important that the skin surface is just touched by the pins. What does that tell you about the relative depth of pressure receptors and pain receptors?

❖ Which area tested is the most sensitive? Explain the advantage of having this area particularly sensitive?

Part III: Ligament and Tendon Reflexes

Reflex responses indicate all parts of the reflex arc are working properly. The reflex arc is a receptor, a sensory nerve, interneurons in the central nervous system, a motor nerve, and an effector (muscle or gland).

1. Patellar reflex. Working in pairs, have the subject sit on the middle of the table (sitting on the edge may break the table) so that the leg from the knee down hangs freely. The examiner should locate the patellar ligament (just below the knee cap or patella), then strike it with the flat edge of a reflex mallet. It is best for the subject to remain relaxed and divert his attention. Notice the degree to which the leg is extended by the contraction of the quadriceps muscle.
   ❖ Test the reflex on right and left legs. Do they respond the same?

2. Achilles jerk. Have the subject kneel on a chair with the feet hanging freely over the edge of the chair. Locate, then tap the tendon of Achilles in the heel of the foot. What is observed?

Name ______________________________ Section ______________

Part IV: Eye Reflexes

1. Corneal reflex. Quickly move the palm of your hand toward your partner's eyes, stopping a safe distance away.
   - ❖ What is observed?
   - ❖ What is the purpose of this reflex?
2. Photo-pupil reflex. Have the subject close his eyes for 2 minutes. Examine the pupil size immediately on opening the eyes.
   - ❖ What is observed?
   - ❖ What is the purpose of this reflex?
3. Convergence reflex. Have the subject look at a distant object (7 or more meters away). Note the position of the eyeballs. Have the subject immediately focus on a pencil held about 25 cm away.
   - ❖ What change occurs in the position of the eyeballs?
   - ❖ What is the purpose of this convergence?

Part V: Swallowing Reflex

1. Swallow the saliva in your mouth and immediately swallow again, then again.
   - ❖ What do you observe?
2. The fact that swallowing can be performed in rapid succession is demonstrated by rapidly drinking a glass of water or some other fluid.
   - ❖ Is swallowing a reflex? Explain.

Name ______________________ Section ______________

Part VI: Olfactory Adaptation (Fatigue Time)

1. Obtain a small bottle of isopropyl alcohol. Close one nostril and smell the solution. Exhale through your mouth. Continue doing this until you no longer smell the alcohol. Record the number of minutes before olfactory fatigue occurred.

2. Open the other nostril.
   - ❖ Can you smell the alcohol?

3. Immediately sniff the bottle of oil of cloves.
   - ❖ Can you smell the cloves?

   - ❖ When olfactory fatigue occurs:
     (1) are all olfactory cells affected or only the cells exposed to the stimulus?

     (2) are the cells fatigued to all chemicals?

Name ______________________ Section ______________

Lab 35

Population Ecology

Developed by Karen Borgstrom

Problem

How do survivorship and mortality rates impact population ecology?

Objectives

After completing this lab exercise, the student will be able to:

1. Construct life tables using basic population ecology equations and preliminary data on a hypothetical population.
2. Construct survivorship curves given population data from a life table.
3. Distinguish between the three basic types of survivorship curves observed among natural populations.
4. Simulate populations and gather data to be tabulated in a life table and plotted on a survivorship curve.
5. Compare survivorship curves of a developed and underdeveloped nation.

Preliminary Information

Populations can be characterized in a number of ways. One way an ecologist can characterize a population is based on mortality rates, or, conversely, survivorship. Since all members of a population are not equal, the life span of each member can and does vary tremendously: some will die shortly after birth and others will survive to be the eldest of a group.

An ecologist can statistically represent the mortality and survivorship of a given population using a life table. Health and life insurance companies have long used life tables to determine rates for members of a particular cohort. A **cohort** consists of a group of individuals in a population who were born at the same time and are researched until the last member of the cohort dies. Among scientists, life tables are used in public health, conservation of endangered species, forestry, management of pests, and ecology. Even governments have an interest in knowing the number of individuals at an age for military service, education, or who might be drawing on Social Security in 2030.

Part I: Construction of a Life Table

Although rather simple to construct, the basic formula used in determining survivorship can be confusing. When using a life table one begins with all members of a cohort just beginning life (i.e., time interval or age is equal to zero). An **'x'** is used to represent a time interval or age, therefore, initially $x = 0$.

Name ____________________ Section ____________

The number of **deaths** during an age group is represented as **'D(x)'** and the number of **survivors at the beginning of an age group** is represented as **'S(x).'** Thus, in Life Table 35.1, S(x = 0) is 200 (i.e., the original number of individuals in the cohort) and D(x = 0) is 20 (i.e., the number of individuals who died during the first age group). Therefore, to calculate S(x = 1) (i.e., the number of individuals alive at the beginning of the second age group) use the following formula:

$$S(x + 1) = S(x) - D(x)$$

Thus, we can determine S(x = 1) by S(0) − D(0) or S(1) = 200 − 20 = 180. As you can see all data can be filled in given the deaths during each age group and using the above equation. Note, the cohort was researched until all members of the population perished.

Table 35.1

Life Table for a Hypothetical Population

x = Age

S(x) = Survivors at beginning of age group

D(x) = Deaths during an age group

x	S(x)	D(x)
0	200	20
1	180	40
2	140	60
3	80	80
4	0	0

A life table for a hypothetical population. The members of this population had a life span of five age groups. Note, x = 0 represents the time from birth to age one and D(x = 0) represents the number of deaths during that time. In this population there were 200 members originally and the last member died between the ages of 3-4.

Another factor used in evaluating a cohort is the mortality rate. Since mortality implies data based on members who are deceased, and we are referring to survivors, we will call this rate the **age specific survivorship or $\ell(x)$**. Age specific survivorship is the *percentage* of individuals from the original cohort who have survived to a particular age. Age specific survivorship can be calculated using the simple formula:

$$\ell(x) = S(x) / S(0)$$

For age 0, $\ell(x)$ is always equal to 1.00 or 100%. We can calculate age specific survivorship for the data from Table 35.1.

Table 35.2

Hypothetical Population Age Specific Survivorship

$\ell(x)$ = Living at beginning of age group

x	S(x)	D(x)	$\ell(x)$
0	200	20	1.00
1	180	40	0.9
2	140	60	0.7
3	80	80	0.4
4	0	0	0

Life table for a hypothetical population, which includes age specific survivorship for data from Table 35.1.

Name ______________________________ Section ______________

Keeping in mind the two formulas used:

$$S(x + 1) = S(x) - D(x) \text{ and } \ell(x) = S(x) / S(0)$$

answer the following questions by completing the life table below.

Table 35.3
Hypothetical Population Age Specific Survivorship

x	S(x)	D(x)	ℓ(x)
0	100	40	1.00
1	60	25	0.60
2	35	15	
3		10	
4		10	
5			

How many members were in the cohort originally? ____________

What is the value of D(x=2)? ____________

What is the value of S(x=3)? ____________

What is the value of ℓ(x=4)? ____________

Part II: Survivorship Curves

Tables are useful for illustrating numerical data, but may require study to illustrate trends. Graphical presentations better illustrate trends, which may be advantageous when studying population ecology. In fact, life table data is often plotted to form a survivorship curve. A **survivorship curve** will plot age specific survivorship, ℓ(x), against a time interval or age of the cohort, x. In addition, a graph will allow one to compare survivorship curves of two or more populations regardless of the original population size.

Figure 35.1 below is a survivorship curve for the data from Table 35.2. **Note,** ℓ(x) is plotted on the y-axis and x is plotted on the x-axis.

Name________________________________ Section______________

Survivorship Curve for Data in Table 35.2.

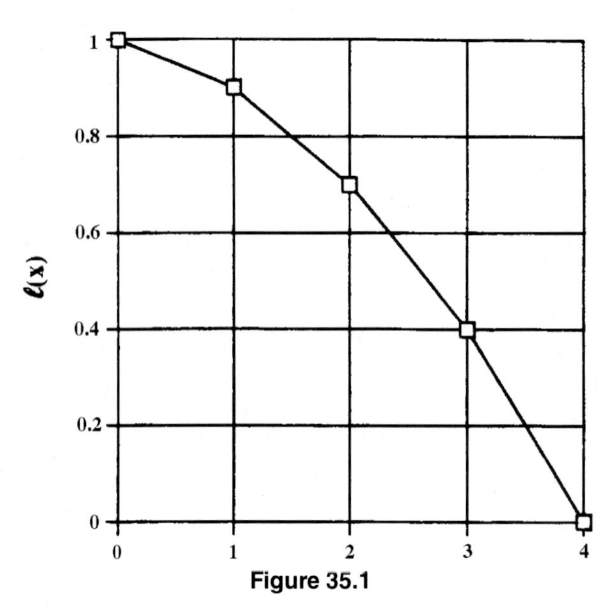

Figure 35.1

Construct a survivorship curve on the graph in Figure 35.2 using the data from Table 35.4.

Table 35.4
Life Table

x	ℓ(x)
0	1.0
1	0.3
2	0.2
3	0.1
4	0

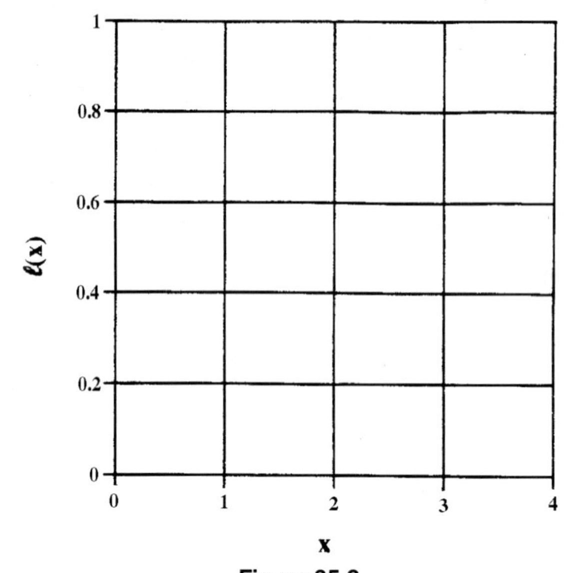

Figure 35.2
Graph of Table 35.4.

Name ______________________________ Section ______________

Part III: The Three Basic Patterns of Survivorship Curves

Below is a description of each of the three basic types of survivorship curves. Keep in mind these curves are based on theoretical data while actual population data may model a certain type but may not exactly fit the curve.

Type I: Late Loss	This type of curve is found in populations that are well adapted to their environment, have few offspring, and provide extensive parental care. In such populations most members live long lives and then mortality abruptly increases very late in life when most members perish. This type of curve is often seen in large mammals, including humans, and annual plants.
Type II: Constant Loss	This type of curve is reflected in populations that have a constant death rate at all ages. Although not very common in nature, this type of curve may be observed in populations of birds, lizards, and small mammals.
Type III: Early Loss	This type of curve is usually seen in populations that have a large number of offspring and little parental care. In such populations there is a high mortality rate early in life. Those members who survive the early years, however, live relatively long lives. Such a curve is seen in fish, invertebrates, and perennial plants.

The graph below shows typical survivorship curves for each of the above survivorship types discussed.

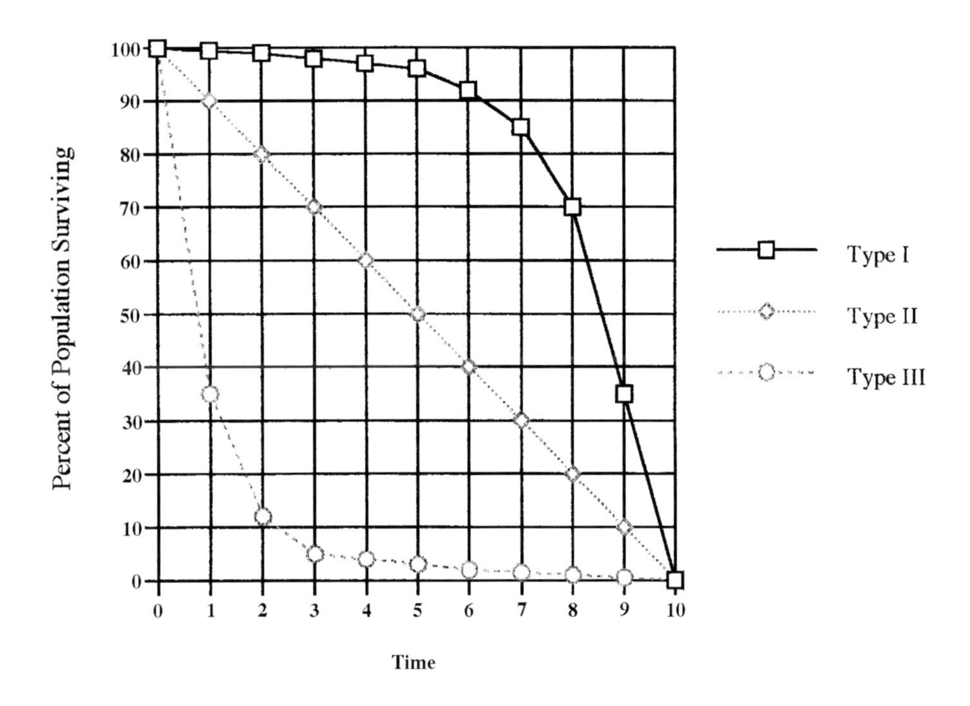

Figure 35.3
Survivorship Curves

Name ____________________ Section ____________

Part IV: Population Simulations

A hypothetical population may be studied using some common everyday objects. Three population simulations will be demonstrated using soap bubbles and dice.

Population Simulation #1

1. Obtain a bottle of soap bubbles.
2. One student will serve as the "bubble blower," one as "timer," and the remaining students will be "bubble counters." The "timer" should count each second out loud so the "bubble counters" know how long the bubble lasted. Each time, the "bubble blower" should blow about three bubbles **high into the air** for each "bubble counter."
3. A total of 50 bubbles should be made and their mortality data recorded in the table for Population Simulation #1. Each bubble represents a single member of the population of 50 soap bubbles. For example, if a bubble is produced and it "pops" at 6 seconds, make a tally mark at interval six in the table.
4. Continue this process until all 50 bubbles have been tallied.
5. Do the calculations necessary to complete the data table for Population Simulation #1.
6. Plot the data x versus $\ell(x)$.
   Note: Data from Population Simulations #1, #2, and #3 are **all plotted on the same graph**. Plot time on the horizontal axis and $\ell(x)$ on the vertical axis. Label each curve.

Notes on Graph Construction

1. A graph is a picture of the data. Plan the graph to properly fit the page; allow for numbers, labels,and margins.
2. A graph should be self-explanatory:
   a. Title must be present
   b. Completely label each axis with factor, units, scale, etc.
3. If several curves are one graph, label each curve.

Population Simulations #2

1. Obtain 50 dice and roll all the dice at once.
2. Each die represents a single member of the population of 50. For a score of 3, 4, 5 and 6 the individual survives time interval one. For a score of 1, the individual dies of heart disease, the die is removed, and a tally mark is recorded in the Heart Disease column. For a score of 2, the individual dies of AIDS, the die is removed, and a tally mark is recorded in the AIDS column.
3. Record the data in the table for Population Simulation #2.
4. Continue to roll the surviving dice and record results until there are less than three dice remaining.
5. Do the calculations necessary to complete the data table for Population Simulation #2.
6. Plot the data x versus $\ell(x)$.

Population Simulation #3

1. Imagine a vaccine for AIDS has been discovered. Thus, repeat the procedure in #2, except for rolls of 2, 3, 4, 5 and 6 the individual survives and for a roll of 1 the individual will die of heart disease.
2. Again, continue to roll the surviving dice until there are less than three remaining.
3. Record data in the table for Population Simulation #3.
4. Do the calculations necessary to complete the data table for Population Simulation #3.
5. Plot the data x versus $\ell(x)$.

Name ______________________________ Section ______________

Table 35.5
Table for Population Simulation #1

Population Simulation #1—Soap Bubble Data				
x (age in sec)	Tally (number dying at each age)	S(x)	D(x)	ℓ(x)
0				
1				
2				
3				
4				
5				
6				
7				
8				
9				
10				
11				
12				
13				
14				
15				
16				
17				
18				
19				
20				

Name ______________________________ Section ______________

Table 35.6
Table for Population Simulation #2

Population Simulation #2—Dice Data					
x	**Tally**		**S(x)**	**D(x)**	**ℓ(x)**
	Heart Disease	**AIDS**			
0					
1					
2					
3					
4					
5					
6					
7					
8					
9					
10					
11					
12					
13					
14					
15					
16					
17					
18					
19					
20					

Name ____________________ Section ____________

Table 35.7

Table for Population Simulation #3

Population Simulation #3—Dice Data			
x	**S(x)**	**D(x)**	**ℓ(x)**
0			
1			
2			
3			
4			
5			
6			
7			
8			
9			
10			
11			
12			
13			
14			
15			
16			
17			
18			
19			
20			

Name ____________________ Section ____________

Part V: Survivorship Curves for the United States and Guatemala

Quality of life conditions may impact mortality rates among human populations. **Plot the data from the Life Tables for the United States and Guatemala [x versus $\ell(x)$] .** Plot both curves on one graph.

Note: If we were to compare survivorship rates of females, they would be slightly higher in both countries due to longer life spans seen in women. However, the difference in survivorship between the United States and Guatemala would be similar in males and females.

Table 35.8

Life Table for the United States and Guatemala

Life Tables for Male Survivorship		
x	**United States— $\ell(x)$**	**Guatemala— $\ell(x)$**
0	1.00	1.00
1	.973	.892
5	.969	.805
10	.966	.779
15	.964	.766
20	.957	.748
25	.949	.728
30	.940	.704
35	.930	.677
40	.916	.648
45	.895	.614
50	.862	.572
55	.812	.523
60	.739	.469
65	.642	.392
70	.518	.310
75	.385	.222
80	.252	.146
85+	.133	.085

Name ______________________________ Section ______________

Part VI: Discussion

1. Which type of survivorship curve do the soap bubbles best approximate? Explain the meaning of this curve.

2. Which type of survivorship curve do the dice in Simulation #3 best approximate? Explain the meaning of this curve.

3. After the elimination of AIDS, is there an increase in the number of people who die from heart disease? Why or why not?

4. Given that the U.S. is a developed nation and Guatemala is a developing nation, discuss some possible reasons for the differences in the two survivorship curves.

Name ______________________________ Section ______________

5. Give examples of organisms and typical characteristics of populations that follow each of these survivorship curves:
   a. Type I: Late Loss.

   b. Type II: Constant Loss.

   c. Type III: Early Loss.

Name ____________ ______________________________ Section ______________

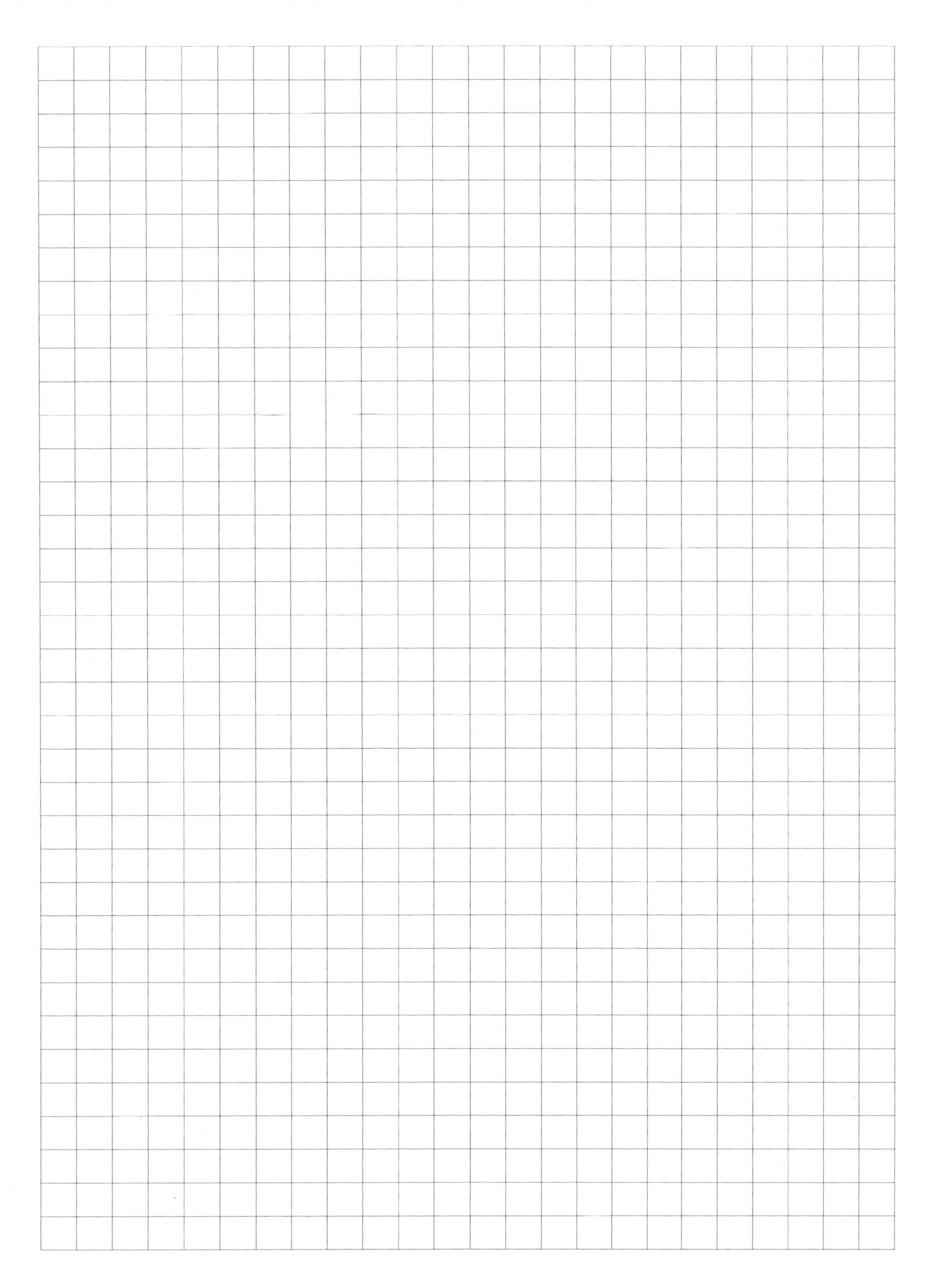

Name ______________________________ Section ______________

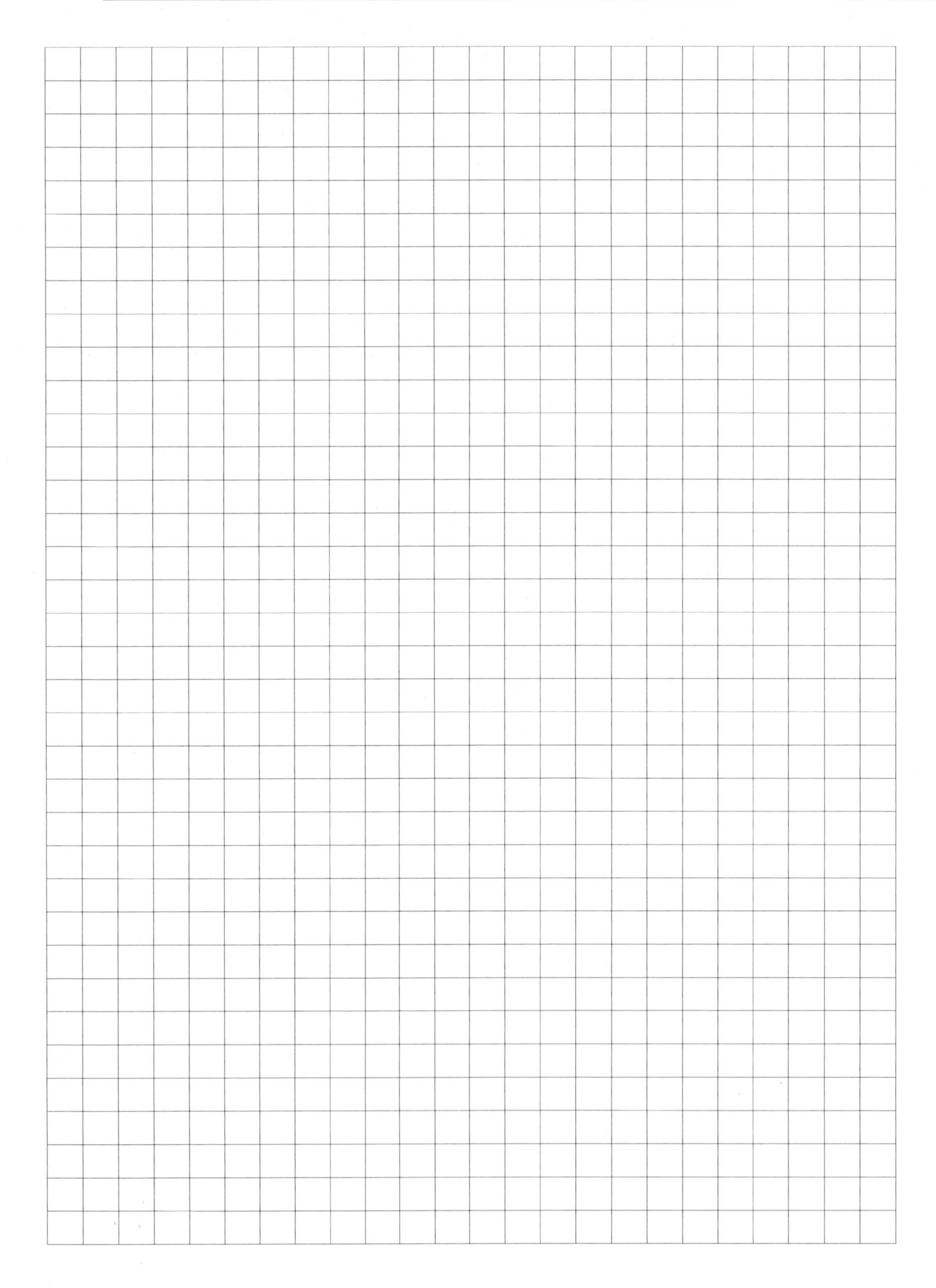

Name ______________________ Section ______________

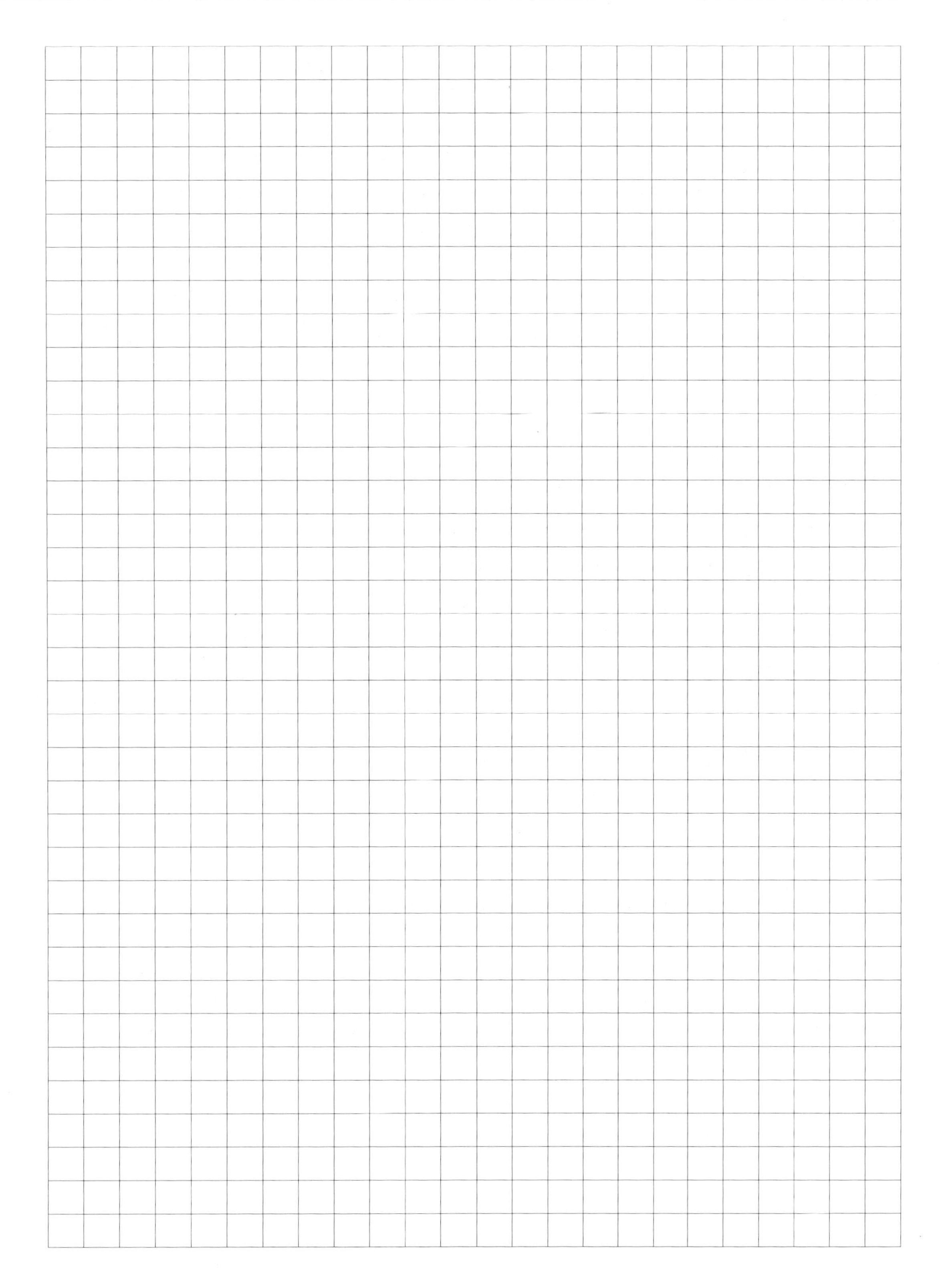

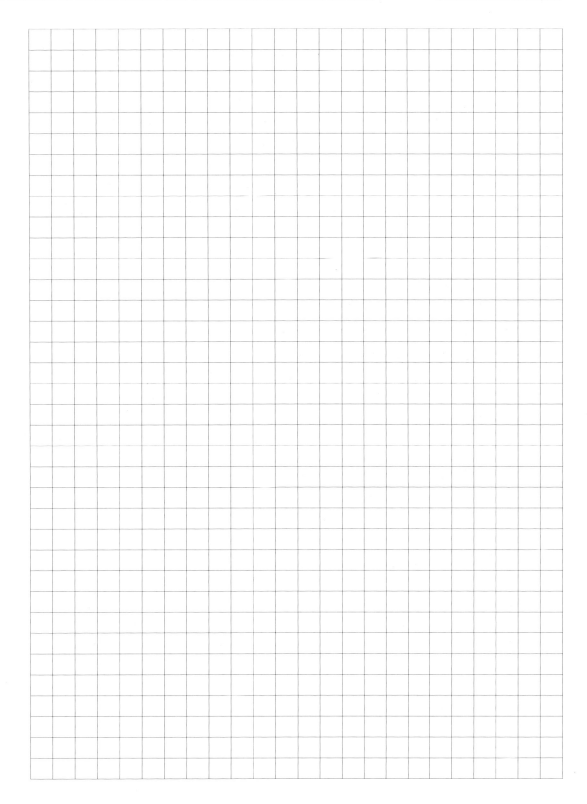

Name ____________________ Section ____________

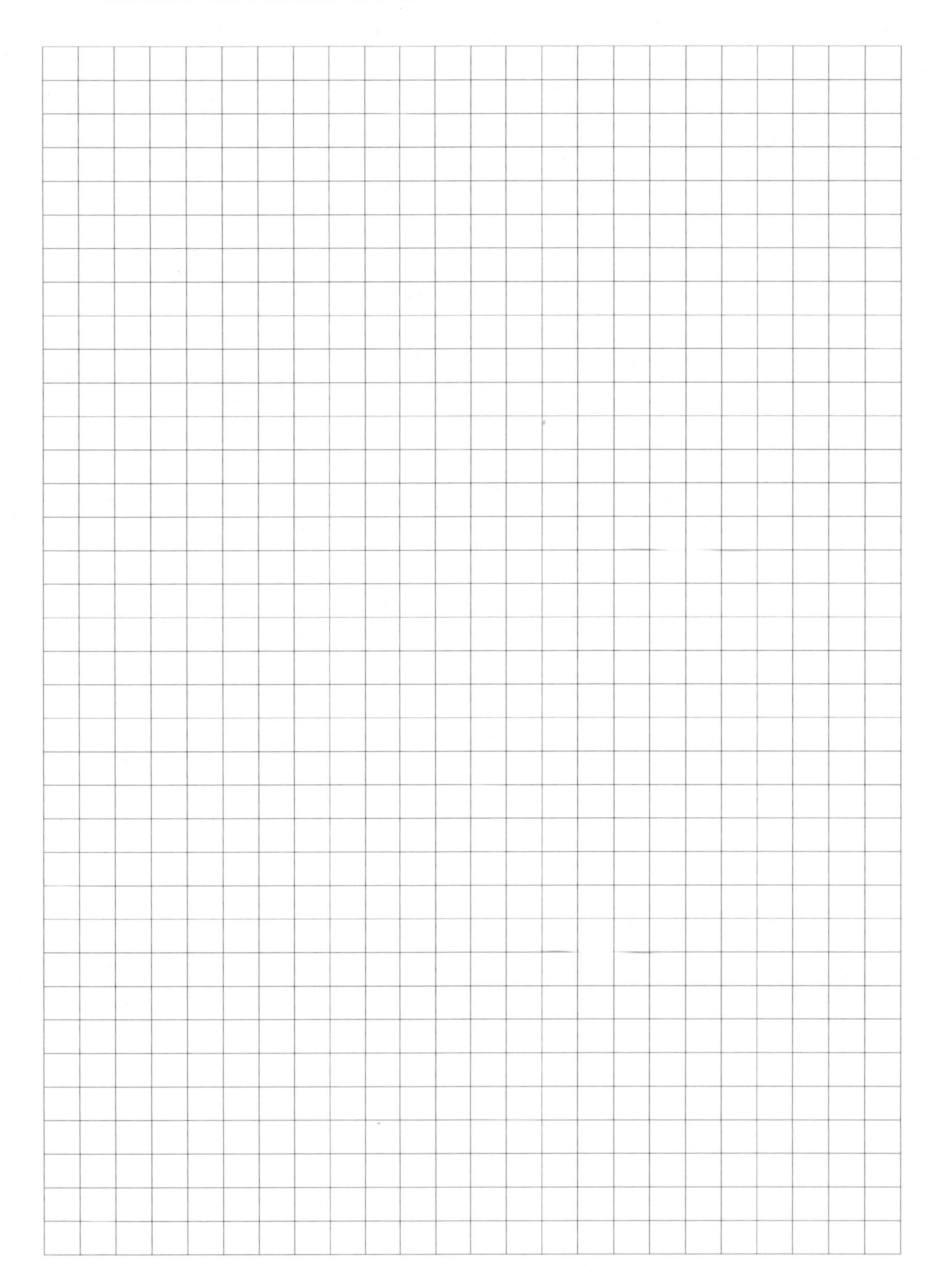

Name______________________________ Section______________

Lab 36

Predation Study of Owl Pellets

Developed by Karen Borgstrom

Problem

What role do owls play in their ecosystem?

Objectives

After completing this lab exercise, the student will be able to:

1. Interpret an energy pyramid.
2. Identify the position of an owl in its ecosystem.
3. Determine the diet of an owl.
4. Diagram a food chain with the owl as the top carnivore.
5. Explain biological magnification.
6. Calculate the daily energy intake of an owl.

Preliminary Information

Predation

Predation is an important interaction between species which helps keep an ecosystem in balance. Because food resources are often limited, predation can also spark competition between members of the same and even different species. Limited resources mean the predator must maximize its food intake with minimal effort, and the prey must maximize its efforts to avoid predation. The enhanced fitness of the more efficient predator and the swifter prey leads to evolutionary changes in both species.

Raptors (carnivorous birds of prey) are top carnivores in the food chain. Hawks and owls are part of the same ecosystem and feed on similar prey but do not compete with each other since the hawks hunt during the day and owls hunt at night. Raptors, such as hawks and owls, catch prey with talons then carry the prey back to their roosts where they will swallow small prey whole or large prey in big pieces. For an owl, the roost tends to be a solitary tree near an open field where the owl hunts. An owl's diet may consist of small mammals, other birds, snakes, frogs, and invertebrates. However, owls cannot digest the entire meal. Indigestible parts such as bones, fur, teeth, feathers and exoskeletons, are formed into tight packages called pellets which are regurgitated about 20 hours after eating. Because the stomach muscles of owls are relatively weak, even the smallest most fragile bones are usually preserved unbroken.

Name ______________________ Section ____________

Food Pyramid

In a terrestrial ecosystem, light energy from the sun is absorbed by plants (the first trophic level) and transformed into plant nutrients such as sugars, proteins and fats. Herbivores (the second trophic level) eat the plants for their nutrients. Primary carnivores (the third trophic level) eat the herbivores. Secondary carnivores (the fourth trophic level) eat the primary carnivores. Thus the nutrients of the plants are passed through a number of trophic levels forming a food chain. The available nutrients and energy at each trophic level decrease because the animals in each trophic level use the nutrients and energy for their own life processes. An average energy use at each trophic level is 90% which means there is a 10% energy transfer to the next trophic level.

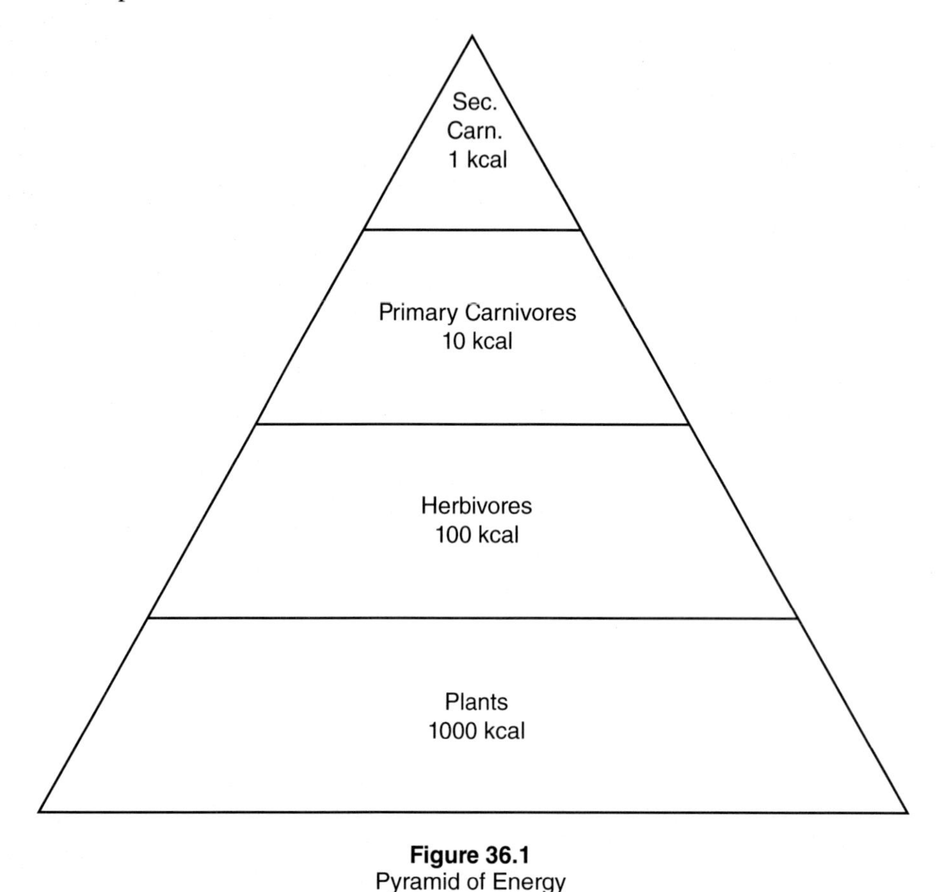

Figure 36.1
Pyramid of Energy

In modern agriculture, plants are often treated with pesticides. Many of these man-made toxins are not biodegradable (such as DDT). The toxins on the plants are consumed by herbivores and passed up the food chain. Since these toxins are not broken down, the toxins are concentrated at higher percentages as they move up the food pyramid. This process is called **biological magnification**.

Owl Pellets

Owl pellets can be collected below the roost site of the owl. Scientists can study owl pellets to determine the diet of the raptor. In this lab, you will be studying owl pellets which have been collected, fumigated and sterilized. Since owl pellets themselves can be tiny ecosystems providing food and shelter for other small organisms, you may find things in your pellet which were not in the diet of the owl. These could include moth larvae, caterpillar droppings (black spheres about the size of a period), and cocoons. Large prey, such as rabbits, may provide a couple of meals for an owl so its remains may be incomplete in a single pellet.

Name ______________________________ Section ______________

Part I: Methods and Materials

1. For each lab group, obtain an owl pellet, 4 forceps, and 5 sheets of white paper, a metric ruler, and a balance.
2. Determine the dry pellet mass of your owl pellet using a balance and measure its length and width. Record this information in Table 36.1.
3. Carefully remove the foil surrounding the pellet. Place the owl pellet on a sheet of white paper. Gently tease apart your owl pellet into large pieces so each group member has a piece.
4. Separate the bones and teeth from the fur, feathers, and other debris.
5. Place any bones and teeth that you find on a separate clean sheet of white paper so they may be more easily seen and studied.
6. Assemble skeletons of the owl prey using the bone sorting chart. Identify the types of prey animals in the owl's diet and how many individuals of each species are present. Record your data in Table 36.2.
7. On a sheet of white paper:
   - Reconstruct the skeletons,
   - Glue the skeletons to the paper,
   - Identify the types of skeletons,
   - Put all group member names on the bottom of the paper.

Part II: Data

1. Use class data to complete Table 36.2.
2. Determine the approximate daily energy intake in kilocalories (kcal) for the owl using class data and this formula:

Daily Energy Intake (kcal) = Mass of pellet (grams) X 35.

Name ____________________ Section ____________

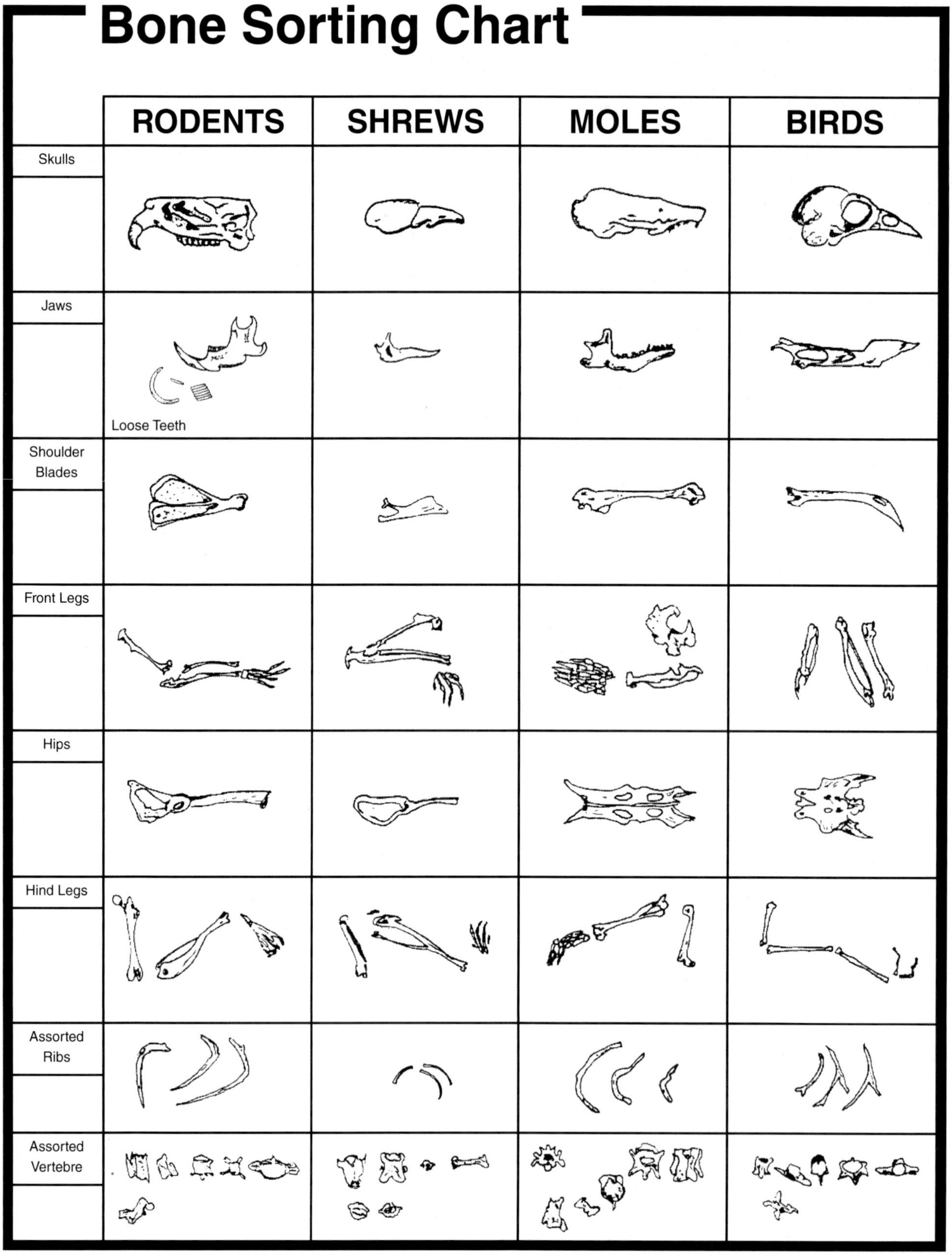

Bone Sorting Chart

	RODENTS	SHREWS	MOLES	BIRDS
Skulls				
Jaws	Loose Teeth			
Shoulder Blades				
Front Legs				
Hips				
Hind Legs				
Assorted Ribs				
Assorted Vertebre				

Figure 36.2
Bone Sorting Chart

Name ____ Section ______________

Table 36.1
Owl Pellet Dimensions

Mass	Length	Width

Table 36.2
Owl Pellet Class Data

Lab Group #	Pellet Mass	Prey Types	# of Organisms of Each Prey Type	Daily Energy Intake in kcal
1				
2				
3				
4				
5				
6				
7				
8				
AVERAGE				

Name ______________________________ Section ______________

Part III: Discussion

1. Assume the average barn owl requires 80 grams of food daily and weighs 480 grams. Calculate the percentage of its body weight that a barn owl must consume on daily basis to survive, using this formula and showing all calculations:

   $$\% \text{ of Body Weight} = \text{Weight of prey} / \text{Weight of owl.}$$

2. Explain why top carnivores, such as birds of prey, are the most threatened organisms in an ecosystem which has been exposed to a non-biodegradable toxin such as DDT?

3. Diagram a food chain with an owl as the top carnivore and the organisms found in its diet placed below it. Beginning with the producers, use arrows to show the flow of nutrients and energy from one organism to another.

Name ______________________________ Section ______________

4. Write the average daily energy intake you calculated from the class data here: ______________ kilocalories. If there is a 10% transfer of energy from one trophic level to another, use your food chain to calculate the number of kilocalories that would be present in the plants that support your food chain. Show all calculations.

5. Competition in ecosystems is reduced when organisms use their habitats in different ways or at different times. This is called resource partitioning. Hawks and owls consume the same prey. How does their hunting behavior differ so that both organisms can exist in the same community without competing with one another?

Name ______________________ Section ______________

Lab 37

Water Ecology Study

Problem

Is there evidence of pollution in water samples collected from the local area?

Objectives

After completing this exercise, the student will be able to:

1. Distinguish between abiotic and biotic factors in the environment.
2. Give an example of a food web found in a local pond.
3. Identify each trophic level of a food chain as a producer, consumer, or decomposer; give an example of each.
4. Determine if the following are at a concentration high enough to be considered aquatic pollutants:
   a. ammonia
   b. coliform bacteria
   c. chloride
   d. nitrate
   e. nitrite
   f. phosphate.
5. Discuss the importance of adequate levels of oxygen and carbon dioxide.
6. Discuss how pH can effect water organisms.
7. Measure the hardness and the pH of a water sample
8. Describe how microorganisms can effect water quality.

Preliminary Information

States bordering on the Great Lakes are fortunate to live by the world's largest supply of fresh water. But concern is growing about increased levels of several pollutants in the drinking water. For example, PCB's or polychlorinated-biphenyl used in industry have entered the food chain. It is recommended that children and pregnant women limit their consumption of fish caught in Lake Michigan to two per year.

Pollution can occur from natural causes, but most water pollution stems from human causes. These range from toxic metals and dyes of industry to silts and acids of mine drainage, from detergents and solid wastes of sewage to potentially dangerous bacteria and chemicals from fertilizers, pesticides, and herbicides. One of the results of accelerated pollution is found in the choking of a lake or stream caused by extensive weed and algae growth. This process, called **eutrophication**, is hastened by increased nitrate and phosphate levels brought on by incomplete waste treatment, or fertilizer run-off from lawns and fields.

Name ______________________________ Section ______________

Every ecosystem has organisms at different feeding levels or **trophic levels**. The **first trophic level** is composed of photosynthetic producers (cyanobacteria, algae, and plants), which make their own food. The **second trophic level** (consumer level I) contains herbivores, which feed on the producers. The **third trophic level** (consumer level II) contains **primary carnivores**, which feed on trophic level II. The **fourth trophic level** (consumer level III) contains **secondary carnivores**, which feed on the third trophic level. **Scavengers** feed on dead organisms from all trophic levels. Decomposers (bacteria and fungi) are essential in recycling elements (such as carbon, hydrogen, oxygen, nitrogen) in the dead producers, dead animals, and waste products. These elements are used as nutients by producers.

These trophic levels can be arranged into a food pyramid with the producers forming the base of the pyramid and each trophic level set in turn on the level below it. Most ecosystems do not have more than four trophic levels because there is not enought energy available to support more than four.

Part I: Materials and Methods

Part IA: Materials

Samples needed for testing can be collected by the students or made available in the laboratory. Student teams of 4 each, going to a pond need to assemble the following items:

1. Water quality kits.
   Note: Each group performs different water quality tests, which are shared with the class.
2. One small jar with lid.
3. One thermometer in a metal case.
4. Plankton net
5. Bucket
6. Pond Life Golden Guide

Part IB: In the Field at a Pond

1. Collect the following samples.
   a. One jar of the surface water. Using the plankton net, collect along the pond's edge among the vegetation. Get the water with as much algae as possible. It will have the best variety of life forms.
   b. Collect a variety of plants growing from the bottom of the pond. Break off a piece several centimeters long. Add pond water to the bucket so the plants and animals in the sample don't die.
2. Measure the temperature of the air and the water.
3. Check water quality using test kits.
4. Using the Pond Life Golden Guide, identify plants and animals seen around the pond. Record and indicate their trophic levels in the data tables.

Part II: In the Lab

1. Complete Biotic Analysis (see instructions later in this lab).
2. Collect class data on all chemical tests.
3. Collect class data on all organisms.
4. Record and indicate the trophic level of each organism in the data table.

Name____ Section________________

Part III: Abiotic Analysis

Using test kits, several physical factors of the water will be measured. You will be assigned specific kits to use. Record data on the data page and share the data with other research groups. Normal ranges are indicated in parentheses.

1. Ammonia (.02-.06 mg/L)

Ammonia is a product of the microbiological decay of animal and plant protein. It can be reused directly to produce plant protein and is used commonly in commercial fertilizers. Ammonia is one source of nitrogen for plants.

Presence of ammonia nitrogen in raw surface water usually indicates domestic pollution. However ammonia nitrogen in ground waters is normally due to natural microbial reduction processes. Its presence in waters used for drinking purposes may require the addition of large amounts of chlorine, which will first react with all the ammonia present to form chloramines.

2. Carbon Dioxide (10 mg/L or less at surface)

Carbon dioxide occurs naturally in water as a product of aerobic or anaerobic decomposition of organic matter; it also is absorbed readily from the atmosphere. Carbon dioxide reacts with water to form carbonic acid. Although the carbon dioxide concentration usually found in water appears to have no physiological effects on humans it has a marked effect on fish and other aquatic life. Continual exposure to concentrations of 100 mg/L or more has been shown to be fatal to many types of freshwater organisms.

3. Chloride (maximum 250 mg/L)

Many rocks contain chloride so its presence in the water may be due to natural processes. Heavy salting of the roads during the winter will increase the amount of chloride in the area as can industrial wastes.

4. Hardness

Hardness represents the concentrations of magnesium and calcium ions. Water passing through soil and rock will dissolve magnesium and calcium. Soft water contains considerable amounts of chloride and sulfate ions, which causes magnesium and calcium ions to precipitate out of solution.

	Amount Dissolved Minerals
Soft water	0-60 mg/L
Moderately hard water	61-120 mg/L
Hard water	121-180 mg/L
Very hard water	over 180 mg/L

Hardness of water does not indicate pollution.

5. Nitrate (maximum 10 mg/L)

Nitrate represents the most completely oxidized state of nitrogen commonly found in water. Nitrate-forming bacteria convert nitrites into nitrates under aerobic conditions and lightning converts large amounts of atmospheric nitrogen (N_2) directly to nitrates. Many granular commercial fertilizers contain nitrogen in the form of nitrates.

High levels of nitrate in water indicate wastes from run off of heavy fertilized fields. Nitrates can degrade water quality by encouraging excess growth of algae. Drinking waters containing excessive amounts of nitrates can cause infant methemoglobinemia (blue babies). For this reason, a level of 10 mg/L nitrate has been established as the maximum allowable concentration of nitrates in public drinking water supplies.

Natural concentrations rarely exceed 10 mg/L and are often less than 1 mg/L.

6. Nitrite (.1 mg/L)

Nitrate nitrogen occurs as an intermediate stage in the biological decomposition of compounds containing organic nitrogen. Nitrite-forming bacteria convert ammonia under aerobic conditions to nitrites. The bacterial reduction of nitrates can also produce nitrites under anaerobic conditions. Nitrites are often used as corrosion inhibitors in industry and cooling towers; the food industry uses nitrite compounds as preservatives.

Nitrites are not often found in surface waters where they are readily oxidized to nitrates. The presence of large quantities of nitrites indicates partially decomposed organic wastes in the water being tested. Drinking water concentrations seldom exceed 0.1 mg/L nitrite.

7. Oxygen (5-12 mg/L)

The amount of dissolved oxygen present in a given volume of water, that is, the solubility of the oxygen, depends both on temperature and atmospheric pressure. **As the temperature increases, the amount of oxygen that water can hold decreases.** As the atmospheric pressure increases, the amount of oxygen that water can hold increases.

Because the oxygen saturation of water is dependent on both temperature and atmospheric pressure, the amount of dissolved oxygen required to make a given volume of water 100% saturated also varies. As determined by testing, at 0°C and 760 mm of pressure (1 atmosphere of pressure), highly oxygenated fresh water will contain approximately 14.16 parts per million (ppm) or 14.16 mg/L of dissolved oxygen; this is said to be 100% saturated. Increase in temperature of the water will cause the oxygen level to decline.

Photosynthesis of aquatic plants and algae produce oxygen. The wave action of ponds and lakes, the shallowness of ponds and streams, and the constant swirling and churning of water over riffles and falls cause a high degree of contact between the water and the atmosphere. It is not surprising, then, that such waters usually come close to being 100% saturated with oxygen.

Generally, 5 mg/L dissolved oxygen content is a borderline concentration for good diversity of life. For adequate game fish population, the dissolved oxygen content should be in the 8-12 mg/ L range.

8. pH (6.0-9.0)

The pH values below 5 (acid) or above 9 (alkaline) are definitely harmful to many animals. In addition, within the normal range, pH can affect the toxicity of poisons. For example, ammonia (NH_3) is more toxic in alkaline water than in acidic water, and cyanides and sulphides are more toxic in acidic water than in alkaline.

The pH also is thought to be a limiting factor for algae and for some invertebrates. It is known that some crustaceans can withstand a wide range of pH values while others are confined to a narrow range.

The pH of many freshwater ponds and lakes has changed due to acid rain. **Acid rain occurs when nitrogen oxides and sulfur oxides in the air combine with rainwater to form acids such as nitric acid and sulfuric acid.** The nitrogen oxides and sulfur oxides are produced in part by the burning of fossil fuel for transportation and electricity.

9. Phosphate (.1 mg/L)

Phosphates enter the water supply from biological wastes and residues, agricultural fertilizer runoff, water treatment, industrial effluents, chemical processing, and the use of detergents contribute significantly.

A certain amount of phosphate is essential to organisms in natural waters and often is the limiting nutrient for growth. Too much phosphate can produce eutrophication or overfertilization of receiving waters, especially if large amounts of nitrates are present. The result is the rapid growth of aquatic vegetation in nuisance quantities, and an eventual lowering of the dissolved oxygen content of the lake or stream due to the death and decay of the aquatic vegetation.

10. Substratum

Bottom type plays a significant role in determining organisms present in a pond or stream and can determine the dissolved mineral content. Sandy bottoms are the least productive as there is little substrate for either protection or attachment. Gravel and rubble bottoms have high productivity. There are large areas for attachment sites and rocks provide abundant nooks and crannies where organisms can hide. Mud bottoms can contain high concentrations of organic materials, which provide nutrients to plants.

11. Temperature

Temperature of rivers and streams vary much more rapidly than those of lakes, but this variation is usually over a smaller temperature range than that of still waters.

The temperature of a stream or lake is very important in determining its species composition. Different species have different temperature minimums and maximums that they can tolerate.

Below are some maximum temperatures tolerated by these various fish:

Carp	36°C–37°C
Perch	30°C
Pike	29°C
Brook Trout	25.3°C
Rainbow Trout	24.5°C

It should be noted that even though a fish can survive at a given temperature, this does not mean that it will have optimal growth or be able to reproduce.

The temperature that fish can withstand depends on such factors as species, age, condition, oxygen content of the water, pH, and the chemical composition of the water.

Temperatures also have a direct affect on the toxicity of poisons to fish. In general, for a given concentration of poison, a rise of 10°C will cut the survival time of fish in half.

Name ______________________________ Section ______________

12. Bacteria

Bacteria are present all over the earth, so their presence in a water sample would be expected. However, if the concentration of bacteria is very high or specific types of bacteria are present, this is an indication of pollution. Coliform bacteria are naturally present in the digestive tracts of vertebrates. High concentrations of coliform bacteria in a body of water can be caused by such factors as large amounts of fecal matter being put directly into the water, run-off from septic fields during heavy rains, run-off from yards with fecal matter from dogs, run-off from feed lots of livestock or poultry, or runoff from landfills.

Part IV: Biotic Analysis in the Lab

1. Microbial Streaking and Incubation

A. Obtain two petri plates: one containing nutrient agar (NA) and a second containing eosin methyl blue (EMB) agar.
B. Label the plates with class section and date in small print on the bottom of the plate.
C. Dip a cotton swab into the water sample and spread the water sample evenly over the entire surface of the NA plate. Immediately close the lid of the petri plate.
D. Then rewet the swab in the pond water and repeat for the EMB plate.
E. After at least 15 minutes, invert each plate and incubate for 24-48 hours at 37°C.

2. Algae and Microscopic Organisms

A. These organisms can range in size from a pin point to huge mats that can cover parts of a pond. Most freshwater algae are green but there are blue-green, yellow, or brown organisms. A cell wall generally is visible.
B. With a pipette, make a few wet mount slides of algae and microscopic organisms. Using the monocular microscope, identify the algae and other organisms.
C. Use the following:
  - *Pond Life* Golden Guide
  - *Freshwater Invertebrates*, Needham
  - *Algae and Water Pollution*, Research Lab, color plates.
D. Record the names, and trophic levels on the data sheet.

3. Larger Organisms

A. Use a dropper to capture larger organisms and transfer them in a single drop to a petri dish for identification using a binocular microscope.
B. Record name, type, and trophic level of each organism.

Name ____________________ Section ____________

Pyramid of Trophic Levels in Freshwater Communities

TROPHIC LEVEL I/
PRODUCERS:
Convert light energy to chemical energy

Single-celled Algae
Filamentous Algae
Photosynthetic protists (phytoplankton)
Vascular Plants

TROPHIC LEVEL II/
CONSUMER LEVEL I:
Feeders on phytoplankton

Protozoa (zooplankton)
Sponges
Rotifers
Tardigrades
Cladocerans
Copepods
Haliplid Beetles
Mosoquito Larva
Clams
Snails

Feeders on larger plants —bryophytes and vascular plants

Nematodes
Crayfish
Mayfly Larva
Snails
Ducks
Muskrats

TROPHIC LEVEL III/
CONSUMER LEVEL II:
Feeders on zooplankton

Planaria
Rotifers
Nematodes
Cladocerans
Copepods
Blackfly Larva
Juvenile Fish

Feeders on larger invertebrates

Dysticid Beetles
Odonata Nymphs
Water Scorpions
Dobson Flies
Water Bugs
Small Fish
Frogs, Salamanders
Turtles
Birds

TROPHIC LEVEL IV/
CONSUMER LEVEL III:
Feeders on larger invertebrates and small fish

Large Carnivorous Fish such as bass and bluegills
Large Frogs
Turtles, Snakes
Birds
Mink

DERITUS FEEDERS:
Feeders on dead organic matter, from all trophic levels

Protozoa
Planaria
Nematodes
Annelids
Rotifers
Clams
Cladocera
Copepods
Ostracods
Amphipods
Blackfly Larva

DECOMPOSERS:
Bacteria
Fungi

Name ____________________ Section ____________

Part V: Data

Pond Study Data Sheet

1. Physical Factors

Table 37.1
Physical Factors of the Pond

Test	Results	Normal Range
1. Ammonia		
2. Carbon Dioxide		
3. Chloride		
4. Hardness		
5. Nitrate		
6. Nitrite		
7. Oxygen		
8. pH		
9. Phosphate		
10. Substratum		
11. Air Temperature		
12. Water Temperature		

2. Microbial Interpretation

A. Nutrient agar is a non-selective culture medium, meaning all bacteria grow on it. Different bacteria often produce different color colonies. Different bacteria can also produce the same color colonies, so the exact type of bacteria cannot be determined by color on the NA plate, but a sample of the variety of bacteria can be determined by analyzing various bacterial colony characteristics. Examine the NA plate. Differentiate the different types of bacteria by colony growth characteristics such as:

- ❖ Color (such as clear, cream, yellow, orange),
- ❖ Shape (round, irregular),
- ❖ Border (smooth, lobed, fringed),
- ❖ Sheen (shiny, dull).

Name ______________________________ Section ______________

Table 37.2
Variety of Pond Bacteria

Species	Approximate Number of Colonies	Color	Shape	Border	Sheen
Bacteria 1					
Bacteria 2					
Bacteria 3					
Bacteria 4					
Bacteria 5					
Bacteria 6					
Bacteria 7					
Bacteria 8					
Bacteria 9					
Bacteria 10					

B. EMB agar inhibits growth of gram-positive bacteria. Bacteria considered detrimental to human health tend to be gram-negative. However, not all gram-negative bacteria are detrimental. Examine the EMB plate. Identify the possible bacteria present using the following chart.

Table 37.3
Gran-negative Bacteria Appearance on EMB agar

Bacteria	Colony Color	Colony Center	Other
Esherichia coli	Pnk	Blue-black	Green-Metallic sheen
Entereobacter	Pink-purple		Mucoid, no sheen
Klebsiella pneumonia	Pink-purple	Dark	Metallic sheen
Proteus	Clear		
Salmonella	Yellow/clear		Translucent
Shigella	Pink/clear		Translucent

List the bacteria that have been identified on the EMB agar.

- ❖ ______________________________
- ❖ ______________________________
- ❖ ______________________________
- ❖ ______________________________
- ❖ ______________________________
- ❖ ______________________________

Name ______________________ Section ______________

Pond Study Data Sheet

Name of Area ______________________ Date ______________

Producers	Trophic Level

Name ______________________ Section __________

Pond Study Data Sheet

Name of Area ______________________ Date __________

Consumers	Trophic Level

Detritus Feeders and Decomposers	Trophic Level

Name ______________________________ Section ______________

Part VI: Discussion

1. What is the relationship between temperature of the water and the amount of dissolved gases in the water?

2. What chemical tests could vary in results between winter and summer? Explain.

3. Did you find any evidence of eutrophication in the pond? Explain.

4. What can you conclude from the data obtained by culturing bacteria on two types of agar?

Name ______________________________ Section ______________

5. Our digestive system contains bacteria that don't harm us, so why should the presence of coliform or other gram-negative bacteria in drinking water be of concern to us?

6. Using the data from the abiotic and biotic examinations of the water samples from a pond, do you consider the pond to be polluted? Explain and support your answer with examples from the data.

7. In underdeveloped countries, people use water that is inadequate or unsafe. From your survey of factors effecting water pollution, what can be done to make the water safe for humans to use? Give realistic and practical solutions, not "clean up the water."

Name ____ Section ________________

8. Why is the pH level of the water important? Why is the pH of many bodies of water in the U.S. and Canada changing?

9a. Which trophic level do you predict to be the largest?

b. Explain the reason for your answer.

c. Do your data support your prediction? Explain.

Name ______________________________ Section ______________

Biology Practical

1.	______	21.	______
2.	______	22.	______
3.	______	23.	______
4.	______	24.	______
5.	______	25.	______
6.	______	26.	______
7.	______	27.	______
8.	______	28.	______
9.	______	29.	______
10.	______	30.	______
11.	______	31.	______
12.	______	32.	______
13.	______	33.	______
14.	______	34.	______
15.	______	35.	______
16.	______	36.	______
17.	______	37.	______
18.	______	38.	______
19.	______	39.	______
20.	______	40.	______

Name ______________________________ Section ______________

Biology Practical

1.	________________	21.	________________
2.	________________	22.	________________
3.	________________	23.	________________
4.	________________	24.	________________
5.	________________	25.	________________
6.	________________	26.	________________
7.	________________	27.	________________
8.	________________	28.	________________
9.	________________	29.	________________
10.	________________	30.	________________
11.	________________	31.	________________
12.	________________	32.	________________
13.	________________	33.	________________
14.	________________	34.	________________
15.	________________	35.	________________
16.	________________	36.	________________
17.	________________	37.	________________
18.	________________	38.	________________
19.	________________	39.	________________
20.	________________	40.	________________

Name ______________________________ Section ______________

Biology Practical

1.	____________	21.	____________
2.	____________	22.	____________
3.	____________	23.	____________
4.	____________	24.	____________
5.	____________	25.	____________
6.	____________	26.	____________
7.	____________	27.	____________
8.	____________	28.	____________
9.	____________	29.	____________
10.	____________	30.	____________
11.	____________	31.	____________
12.	____________	32.	____________
13.	____________	33.	____________
14.	____________	34.	____________
15.	____________	35.	____________
16.	____________	36.	____________
17.	____________	37.	____________
18.	____________	38.	____________
19.	____________	39.	____________
20.	____________	40.	____________

Name ______________________________ Section ______________

Biology Practical

1.	______________	21.	______________
2.	______________	22.	______________
3.	______________	23.	______________
4.	______________	24.	______________
5.	______________	25.	______________
6.	______________	26.	______________
7.	______________	27.	______________
8.	______________	28.	______________
9.	______________	29.	______________
10.	______________	30.	______________
11.	______________	31.	______________
12.	______________	32.	______________
13.	______________	33.	______________
14.	______________	34.	______________
15.	______________	35.	______________
16.	______________	36.	______________
17.	______________	37.	______________
18.	______________	38.	______________
19.	______________	39.	______________
20.	______________	40.	______________

Name ____________________ Section ____________

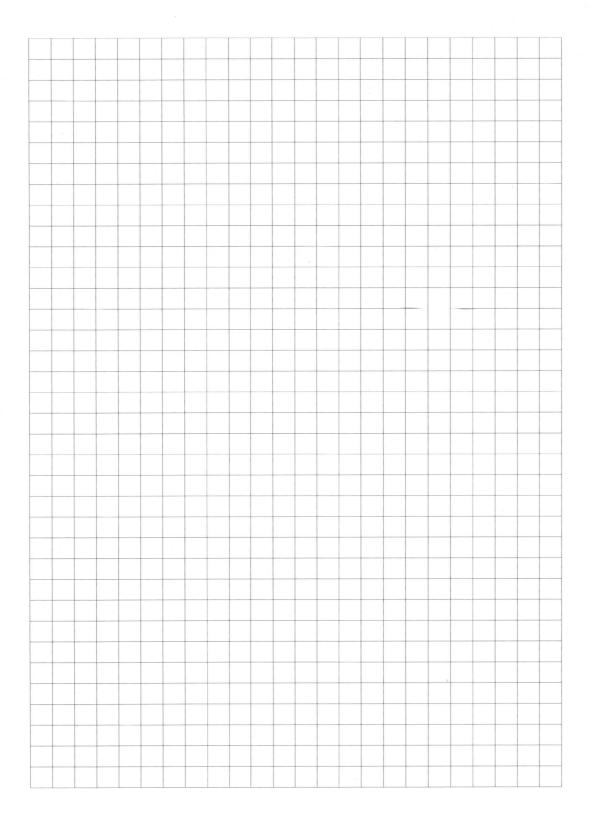

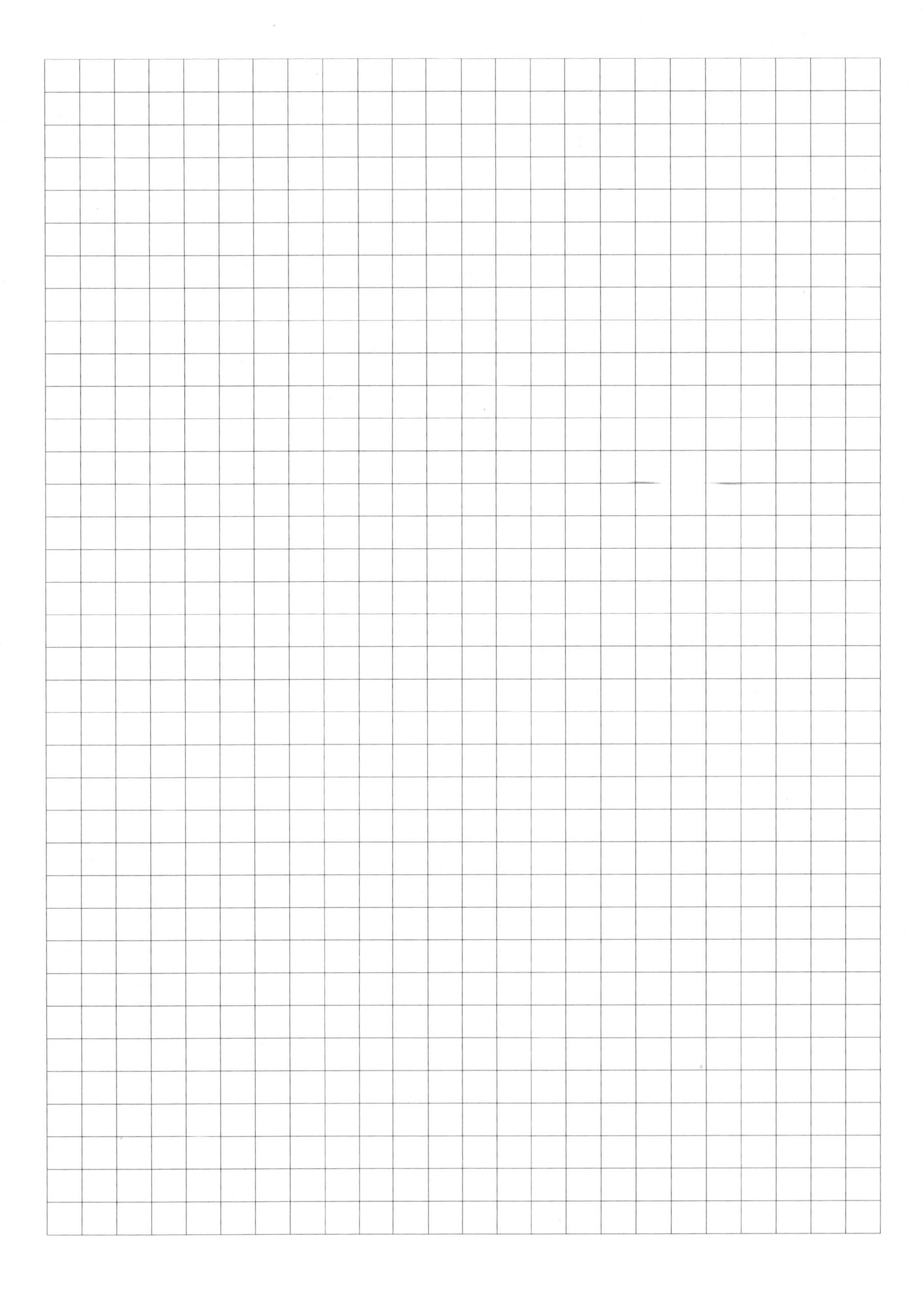

Name ______________________________ Section ______________

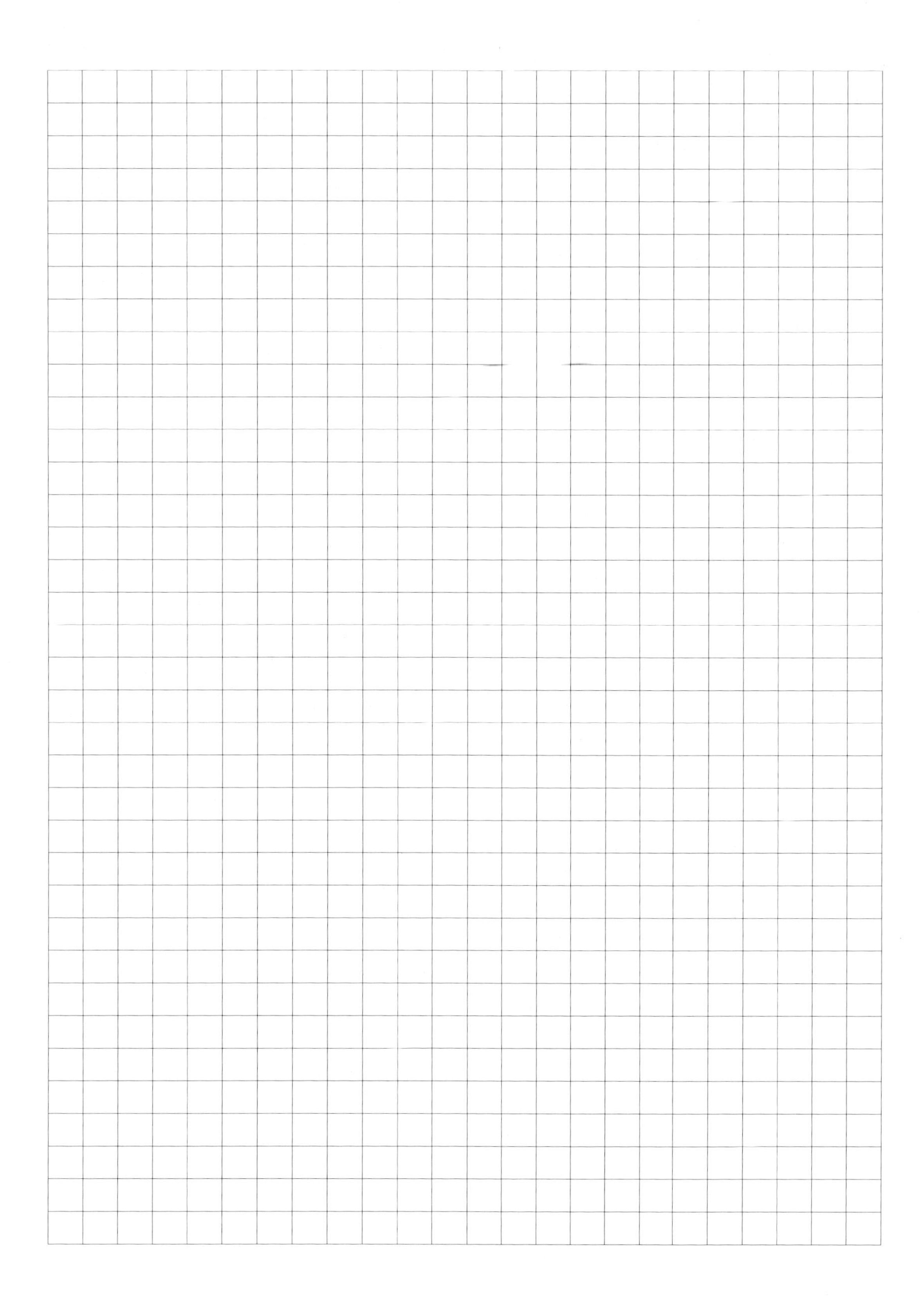

Name ______________________________ Section ______________

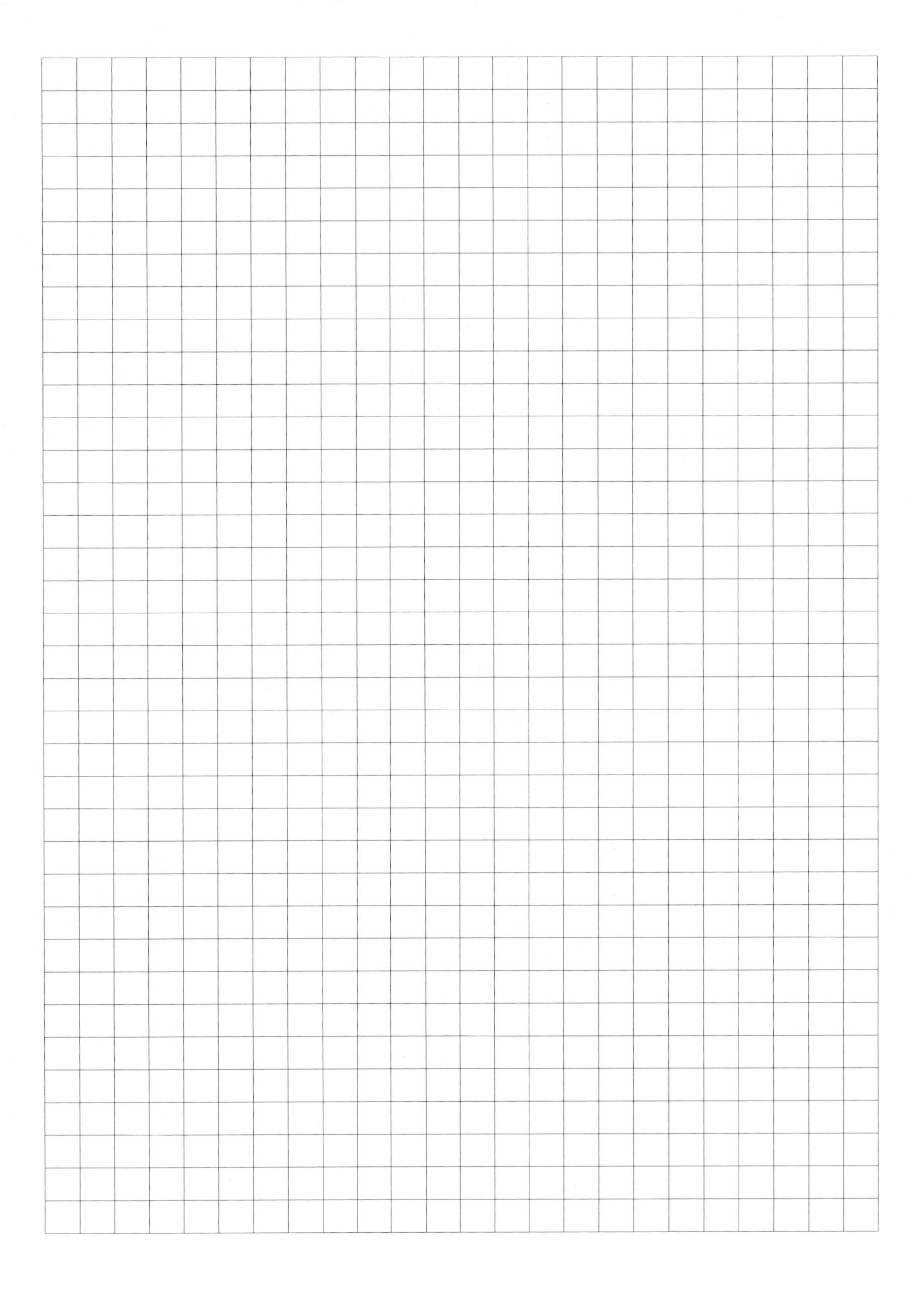

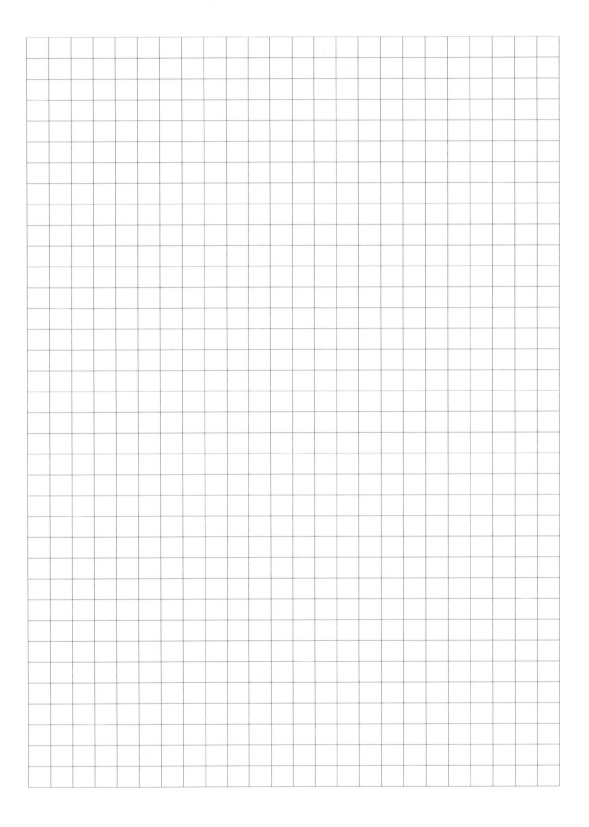

Name ____________________ Section ____________

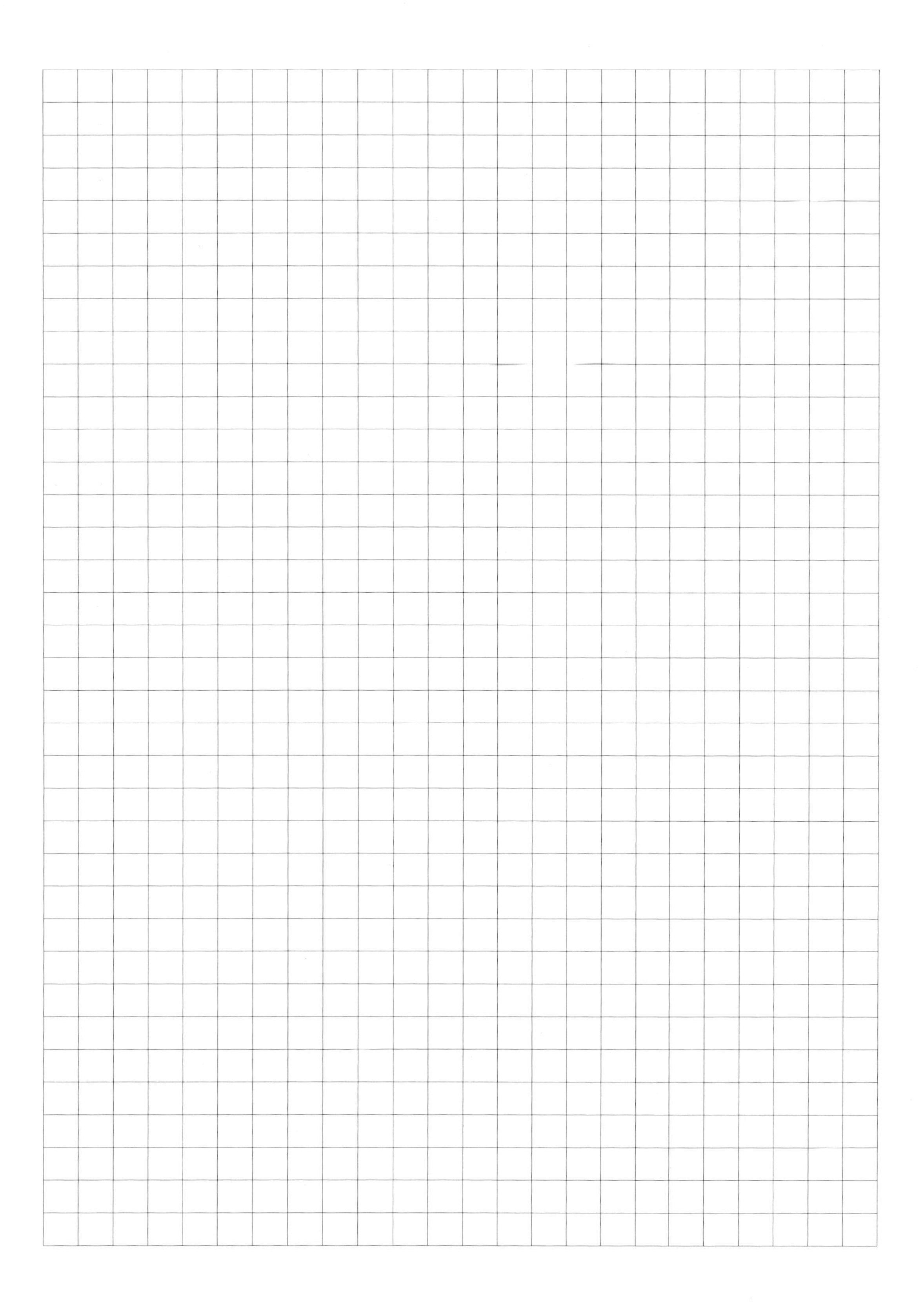

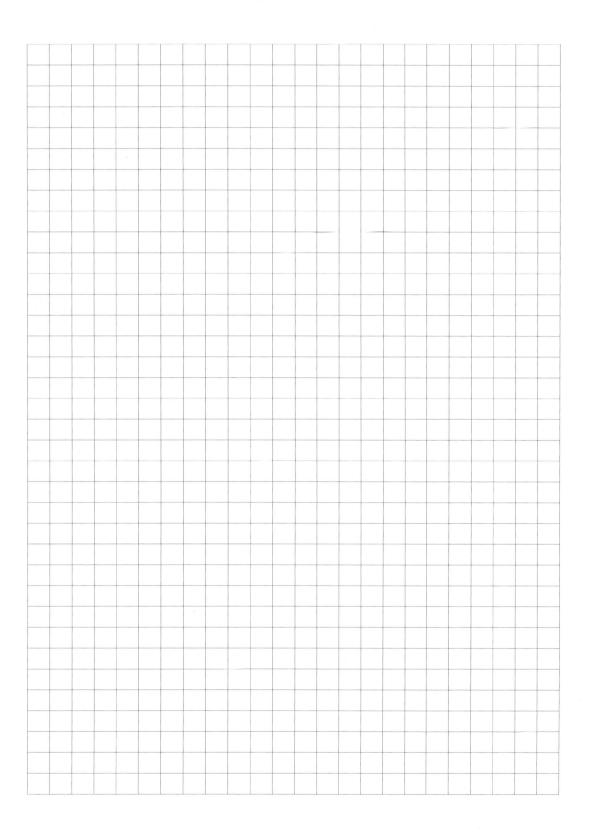